RESTORE THE FUTURE

RESTORE THE FUTURE

The Second American Revolution

Donald H. Young

iUniverse, Inc.
Bloomington

Restore the Future
The Second American Revolution

Copyright © 2012 by Donald H. Young

All rights reserved. No part of this book may be used or reproduced by any means, graphic, electronic, or mechanical, including photocopying, recording, taping or by any information storage retrieval system without the written permission of the publisher except in the case of brief quotations embodied in critical articles and reviews.

iUniverse books may be ordered through booksellers or by contacting:

iUniverse
1663 Liberty Drive
Bloomington, IN 47403
www.iuniverse.com
1-800-Authors (1-800-288-4677)

Because of the dynamic nature of the Internet, any web addresses or links contained in this book may have changed since publication and may no longer be valid. The views expressed in this work are solely those of the author and do not necessarily reflect the views of the publisher, and the publisher hereby disclaims any responsibility for them.

Any people depicted in stock imagery provided by Thinkstock are models, and such images are being used for illustrative purposes only.

Certain stock imagery © Thinkstock.

ISBN: 978-1-4620-8385-5 (sc)
ISBN: 978-1-4620-8384-8 (hc)
ISBN: 978-1-4620-8383-1 (e)

Library of Congress Control Number: 2011963234

Printed in the United States of America

iUniverse rev. date: 01/23/2012

This book is dedicated to the following members of my family – my wife, Karen, who encouraged me to write this book; our sons, Rob and Ryan, of whom we will always be extremely proud and who represent our enduring legacy; our grandchildren; and my parents, Ellen and Don Young, who provided tremendous educational and developmental opportunities for me.

It is also dedicated to all of you. I love this country and the opportunities it has provided for me, but I am extremely concerned that these opportunities will not exist for our children and grandchildren. After reading this book, I hope you will join me in the enormous but critical effort to Restore the Future through the Second American Revolution.

TABLE OF CONTENTS

	Introduction	1
Chapter 1	The Role of Government	5
Chapter 2	The Indictment and Call to Revolution	17
Chapter 3	Morality and Religion	49
Chapter 4	Limited Government	75
Chapter 5	On Liberty	101
Chapter 6	The Rule of Law	120
Chapter 7	National Defense	140
Chapter 8	Education	158
Chapter 9	Free Enterprise	186
Chapter 10	Conclusion	211
	References	217
	Index	227

INTRODUCTION

The budget should be balanced, the Treasury should be refilled, public debt should be reduced, the arrogance of officialdom should be tempered and controlled, and assistance to foreign lands should be curtailed lest Rome become bankrupt. People must again learn to work, instead of living on public assistance.

> Cicero
> Roman philosopher – 55 BC

If you know the enemy and know yourself, you need not fear the result of a hundred battles. If you know yourself but not the enemy, for every victory gained you will also suffer a defeat. If you know neither the enemy nor yourself, you will succumb in every battle.

> Sun Tzu
> Chinese military general and strategist

For the first time in American history, a self-induced shadow has fallen over the American future. Whether this is a permanent change or something like an eclipse will depend on the policies we adopt over the next few years.

The economic growth potential for this country is being revised down by economists. Relatively high rates of unemployment may become the new normal, because economic growth is so anemic. Only half the people pay taxes. Competition is increasing on a global basis. A long-term decline in the value of the dollar may be under way. The people are deeply divided over the future direction of the country, and the resulting malaise is tremendously counter-productive to a dynamic future. Our position as a global super power may be slipping.

In this environment, it may be increasingly difficult for our children and grandchildren to achieve their dreams.

The purpose of this book is to galvanize the American people to "Restore the Future" by becoming participants in a Second American Revolution (non-violent). The objectives of this Revolution are to:

- Restore the Constitution to its critical role in our society as one of our key founding documents.
- Roll back all the changes in our government that have been justified by the view that the Constitution is a "living document".
- Revitalize the American people to demand a future for their children and grandchildren which will both inspire and reward them
- Bring back the hope and opportunity which have always characterized American society.

Chapter 1 provides a brief review of forms of government and makes the point that our founding documents created the pinnacle of the development of government at that time. Nothing that has happened in this country or around the world has improved on that form of government and the rationale for its creation which were put in place over 220 years ago.

Chapter 2 follows the outline of the Declaration of Independence in presenting a long list of grievances which demonstrate how far we have moved away from the founding documents and providing the reasons why calling for a Second Revolution which will restore them to their central role is so vital at this time.

Chapter 3 discusses religion, the elemental morality it provides, and the importance of morality to the necessary function of society. The beliefs of the Founding Fathers were that: (1) a superior being exists, (2) natural rights uniquely come from the superior being, and (3) a government must be created with the primary purpose of protecting these rights.

Chapter 4 provides an analysis of the extent to which the limited government created by the Founders is alive and well today.

Chapter 5 discusses the essential nature of liberty, how it is different from a related concept called freedom, and why expanding liberty should be one of the few primary responsibilities of government.

Chapter 6 outlines the derivation and the importance of the Rule of Law to the fundamental functioning of a society.

Chapter 7 discusses the critical role of national defense and why, like expanding liberty, national defense should be one of the few primary responsibilities of government.

Chapter 8 reviews the key role played by education in maintaining the position of this country in the world and the risks we are running as a result of the poor quality of our education relative to that of the rest of the world.

Chapter 9 outlines the rationale for and critical importance of free enterprise and free trade and why free enterprise is the greatest engine for economic growth and development ever created.

Chapter 10 outlines what each citizen should do to enlist in the Second American Revolution and "Restore the Future".

As the quotation from Cicero at the beginning of the Introduction demonstrates, this is not the first time someone has asked for a return to the basic elements of good government. The difference for us 2,066 years later is that we have founding documents which provide the structure for the best government ever created by man, and we should demand that those documents and that government be reinstated.

Donald H. Young

The quotation from Sun Tzu emphasizes the importance of knowing your enemy and knowing yourself and thus greatly increasing your chance for success in this Revolution. This book will help you do both.

CHAPTER 1
THE ROLE OF GOVERNMENT

That Government is Best Which Governs Least

Henry David Thoreau

Background

In principle, men and women have two choices in the way they live their lives. Either they live by themselves or in independent family groups, or they live as a part of a larger society. If they live by themselves, they must provide from their own resources the food they eat, whatever medical care they require, the clothes they wear, the dwellings in which they live, and their defense against predators. They are entirely free, and they have no obligations to anyone other than themselves or their families. This is a formidable list of individual responsibilities, and it is not clear that many people have lived this way.

In fact, many philosophers doubt that they have. As David Hume says in his <u>A Treatise of Human Nature</u> (1739), human beings are naturally social: "'Tis utterly impossible for men to remain any considerable time in that savage condition, which precedes society; but that his very first state and situation may justly be esteem'd social. This, however, hinders not, but that philosophers may, if they please, extend their reasoning to the suppos'd state of nature; provided they allow it to be a mere philosophical fiction, which never had, and never cou'd have any reality."[1]

If people live as part of a society, there are economies of scale in food production, because food is generally produced communally, and there is a joint commitment to national defense for protection of the whole society. In any society, people give up some of the absolute freedoms they might have had in what the French philosopher Jean Jacques Rousseau called a "state of nature", but they gain more liberty to pursue more fully their individual interests. The difference between freedom and liberty is discussed in Chapter 5.

To expand their liberty, people decide to participate in what amounts to a contract with the society. They commit to participate in the society in certain ways, and they expect certain things in return. Rousseau called this the "Social Contract".

"The heart of the idea of the social contract may be stated simply: Each of us puts his person and all his power in common under the supreme direction of the general will, and, in our corporate capacity, we receive each member as an indivisible part of the whole."[2]

One of the best examples of a social contract is the Mayflower Compact. There were two distinct groups on the Mayflower, and there were sharp differences of opinion between the groups about the form of government which should be established once they landed in the New World and began to set up a society. They agreed to a set of rules while still on the ship, which they called a "Compact". They each gave up some of the things they might have enjoyed if they had decided to live in separate groups in the interest of establishing a durable and successful society which provided much more liberty.

It is one thing to recognize that people live in societies, but it is obvious to everyone that the actual creation of societies in recorded history has produced a broad range of outcomes. In primitive societies, there was a chief, who had all the power, and his followers. Sadly, this approach still exists today in parts of the world. In ancient Greece, there were attempts to create a "pure" democracy, in which whatever was done was the result of the decision of a simple majority of all the citizens. This proved to be unworkable and impractical when the population grew beyond a certain point.

There have been dictators and tyrants. There have been hereditary monarchies. There are theocracies, in which the leaders are leaders because religion puts them there. There have been socialism and communism. There have even been combinations of these separate approaches. As varied as the actual societies have turned out to be, they are still societies, in the sense that they are all alternatives to the state of nature.

By definition, societies to function at all must have certain rules, or laws, to establish both how men should conduct themselves to remain a member of the society and what happens to them if they don't follow the rules. The laws are fundamental to the concept of "society".

The principal differences among all of these societies are in the laws by which they function. In primitive societies, the laws are established by the alpha male, as is the case with many groups of animals. In others, the rules are established by whoever has the military power. In many modern societies, the rules are established by the people in the society themselves and not imposed on them. To be effective for any society to function, the rules must be clear, and they must be enforced.

The development of concepts of individual rights and laws to protect those rights for those who participate in a social contract has been evolutionary and not without setbacks and disruption. What we can see is that the leading ideas about these concepts today are clearly not those of the American Indians, the Egyptians, Attila the Hun, or, for that matter, the Nazis or the Russians. This evolutionary process has never been driven by a motivation to reject the social contract and try something else, but rather by a motivation to make it represent better the original purpose for which it is created, as described above.

By the eighteenth century, concepts of individual rights were being developed, and primitive laws to protect them were being established. At the time, the best thinking in what today would be called the western world was that the rights came from the monarch. In some indirect sense, they may have come from a supreme being, but the monarch, in any case, was the one who translated the divine intention. This situation was commonly referred to as the "divine right of kings".

The American Revolution, with the philosophical background largely established by a group of Scottish philosophers, changed all that. John Locke, for example, in his Second Treatise on Government sounded the charge. His view was that:

- All men are endowed by a supreme being with natural rights which are an integral part of the fact that they are humans. Regardless of the society in which one finds himself, these rights are still the same and are not defined by that society. These rights include life, liberty, and property.
- Governments are instituted among men for the sole purpose of protecting and advancing those rights.
- If governments fail to do so, they can be, and possibly should be, overthrown.

Here is what Thomas Jefferson said in the Declaration of Independence, which is a direct reflection of the view of Locke:

"We hold these truths to be self-evident, that all men are created equal, that they are endowed by their Creator with certain unalienable Rights, that among these are Life, Liberty and the pursuit of Happiness [John Locke's "property"] – That to secure these rights, Governments are instituted among Men, deriving their just powers from the consent of the governed, - that whenever any form of Government becomes destructive of these ends, it is the Right of the People to alter or to abolish it, and to institute new Government, laying its foundation on such principles and organizing its powers in such form, as to them shall seem most likely to effect their Safety and Happiness".

The rules for organizing society and a government took a great leap forward with the signing of the Declaration of Independence and the creation of the Constitution. In addition to the requirements indicated above, of recognizing natural rights and creating a government to enforce them, the rules for enforcing them now were designed to be created by the people and to be productive of their collective interests. To make sure that this happens in fact as well as on paper, the Founders created a representative democracy.

The Constitution by itself does not address what the individual rights are. These are expressed in the Bill of Rights, which consists of the first ten amendments to the Constitution and which was passed into law on 12/15/1791.

James Madison said in Federalist #50, "If men were angels, no government would be necessary. If angels were to govern men, neither external nor internal controls on government would be necessary. In framing a government which is to be administered by men over men, the great difficulty lies in this: you must first enable the government to control the governed; and in the next place oblige it to control itself. A dependence on the people is, no doubt, the primary control on the government; but experience has taught mankind the necessity of auxiliary precautions".

The Founders had this famous saying in mind when they constructed a government with appropriate checks and balances, limited and specifically enumerated powers for the federal government, and reliance on the people for its on-going authority. The government was subordinate to the people and not the other way around. The representatives elected by the people worked for the people; the people did not work for the representatives.

In my opinion, this is the pinnacle so far in the evolution of the role of government in a society. No alternative approach has been developed anywhere in the world in more than 220 years which comes close to the comprehensiveness, effectiveness, and legitimacy of the American approach. Many other approaches have been tried and failed – Nazism, communism, socialism, and totalitarianism have all failed to meet the standard set by the founding documents. In fact, many of the emerging nations around the world today are struggling toward the American approach and trying not to reinvent the wheel by pursuing some of the approaches to establishing a government which have already failed.

It is irrelevant to suggest that the Constitution and the Bill of Rights were created by white men, some of whom were slave owners, and did not address certain problems, including slavery and universal suffrage. As most readers of American history know, without the southern states there would not have been a successful revolution. The southern states would not have joined the new government if the

Constitution had tried to address the issue of slavery directly. In fact, agreement was reached to end slavery in 1815, although it clearly wasn't effective.

The important point is that we have a unique country today not only because of what the Founders did at the time, but because they put in place a philosophical framework which allowed the issues of slavery and universal suffrage to be addressed eventually in the context of the Constitution which they created. Do you think for a moment that these issues were addressed by appeals to anything else? Isn't it true that the philosophical foundation for addressing these and other problems was already in place?

Why is this the pinnacle? Because the basic ideas around which the American Constitution, the Bill of Rights, and our form of government were constructed were totally unprecedented at that time in world history and are as valid today as they were 220 years ago.

- The rights of individuals
- The liberty to exercise those rights
- Freedom of speech
- Freedom of religion
- Government of the people, by the people, and for the people
- Limited federal government with carefully enumerated powers
- Commitment to strong national defense
- Freedom of the press
- The right to a trial by a jury of one's peers with the presumption of innocence
- Property rights

The approach we have developed does make America a special place, and, as a result, people of all races and nationalities are still flocking to this country to live and work and to ultimately become American citizens. However, I have to stress here that while our approach has set the standard, our system of government is not unique. We don't have a monopoly on it, and any nation can imitate it and try to improve it whenever it wants.

Since, as described above, changes in the role of government in societies have been evolutionary, it would be presumptuous to believe that our state-of-the art in government today cannot be improved. At this point, however, it is not clear what the evolutionary improvements would be, and it is quite clear in any case that they are not in evidence at this point.

Not only has no other nation succeeded in raising the bar by improving our system of government, but we have not in a fundamental sense done so either. It is truly remarkable that 220 years after its creation we are still trying to follow its basic principles and design.

We have a tripartite government, separation of powers, a bicameral legislature, free elections of the President and our representatives, changes in government based on specified procedures and unaccompanied by violence, and trial by a jury of one's peers with the presumption of innocence, all very much as the Founders designed them. Even in rare situations, such as the impeachment of a sitting president, we follow the rules created in the Constitution to the letter.

In short, our Constitution has stood the test of time, which is a real tribute to the remarkable creativity and vision of the Founders. It should be, as a result, the obligation of every citizen to learn about and vigorously defend the founding documents and the life they have allowed the American people to create. It seems to me that one of the prescriptions for a decline in any society is a lack of willingness to work hard to defend what it has created and why. Indifference and societal malaise are deadly to the preservation and expansion of liberty.

What has changed over two centuries is the implementation of the principles of the Constitution. Benjamin Franklin is reported to have been approached by a Mrs. Powel of Philadelphia on the street in Philadelphia at the end of the Constitutional Convention in 1787 who asked, "Well, doctor, what have we got, a republic or a monarchy?" With no hesitation whatsoever, Franklin responded, "A republic, if you can keep it".[3]

In Chapter 2, I will review how much we have departed from the Constitution and how we are locked in a life-and-death struggle to keep our country consistent with the Constitution and the views of the Founding Fathers and to avoid "losing it". Unfortunately, there is at the beginning of the 21st century a very strong trend in the direction of marginalizing the Constitution by making it a "living document", something which has been developing for more than 100 years. In this view, the Constitution is simply what people want it to be at any point in time.

The Founders, of course, thought of the need to revise and update the Constitution from time-to-time, and they devised a rigorous process for amending the Constitution. That this process works is demonstrated by the fact that there are now 27 amendments, the last of which was ratified in 1992.

To think of the Constitution as a "living document" is to violate the Constitution by amending it without following the formal process created for amendments in the first place. If that can be done even once, it makes a mockery of the Constitution and the carefully designed process for keeping it up-to-date. In this case, there are then no limits on the extent to which the Constitution can be informally amended, based on arbitrary judgments and party politics.

What support of a "living document" really means is that more and more people are thinking that the Constitution must be adapted to fit the current circumstances, because they are not aware of or willing to live up to its provisions. What should be happening is just the opposite. We should be adapting current circumstances so that they are in line with the provisions of the Constitution. If the current trend continues and we are not successful in arresting it, the Constitution will be vitiated, and we will have failed to "keep" our country – the possibility raised by Franklin.

The Honorable Clarence Thomas, Supreme Court Justice, was recently asked the following question at a presentation he made about a book he had written: "What worries you most about the future of the country?" His answer, which echoes Franklin, was "ignorance of the Constitution".

This concern is not new. Alexander Hamilton said, "If it be asked, "What is the most sacred duty and the greatest source of our security in a Republic? The answer would be, An inviolable respect for the Constitution and Laws".[4] Thomas

Jefferson in a famous quote said, "If a nation expects to be ignorant and free, in a state of civilization, it expects what never was and never will be."[5] The Founders were looking down the long span of future history and warning the American people of what was needed to preserve what they created.

One of the biggest threats to the primacy of Constitutional principles has been the existence of the Progressives. Progressives have been on and off the American political stage for 100 years, and their influence probably peaked during the time when President Franklin Roosevelt tried to pack the Supreme Court with justices who agreed with his Progressive political philosophy in the 1930s.

"The Progressives were the self-conscious social and legal reformers who occupied center stage in the period roughly from the onset of the 20th century through the election of Franklin Delano Roosevelt as president in 1932... Progressives believed in the power of science and economics, employed by government, to lift up the economic and social position of the general population"[6]

There was only one problem. Several provisions of the Constitution stood in the way. As a result, "The Progessive legal reformers readily swept aside the "parchment barriers" imposed by the Constitution. The document, they said, must not be interpreted in a static, formal way. Rather than attempt to derive formal rules from the text, the Court should take account of sociological and economic studies of current conditions."[7]

The bottom line is that by "drawing on the new social sciences and a primitive understanding of economic relationships, their efforts reached fruition during the New Deal when the Constitution was essentially rewritten, without the benefit of amendment."[8]

"The Progressive movement did indeed repudiate the principles of individual liberty and limited government that were the basis of the American republic."[9]

The thinking of progressives is dramatically illustrated by Frank J. Goodnow, the [then] president of Johns Hopkins University, who "...noted approvingly (in a 1916 lecture) that in Europe, unlike in America, the rights an individual possesses are, it is believed, conferred upon him, not by his Creator, but rather by the society to which he belongs. What they are is to be determined by the legislative authority in view of the needs of that society. Social expediency, rather than natural right, is thus to determine the sphere of individual freedom of action."[10] The philosophy described in this lecture is so diametrically opposed to 'endowed by their Creator" in the Declaration of Independence that it boggles the mind. No longer are rights "natural" coming from God, they are rather to be determined by society. This is complete inversion of the principles of the American founding, and it shows how dangerous to our country and to liberty Progressivism is. Please remember this quotation when you read Chapter 3 – "Religion and Morality".

Progressivism is insidious in that this particular brand of elitism can be foisted off on the people of this country without them even knowing about it. How many people in the US would support the Progressive view described above, as opposed to the philosophy of the Founders, if they understood the differences? I believe that the percentage would be remarkably small.

Progressives have been responsible for an enormous negative impact on the effort to maintain and defend the Constitution, and the important thing to note is that few, if any, of the major changes in interpretation which they have created and which have led to the kind of big government solution to every problem we have today have been rolled back.

Examples include: (1) Social Security, which was originally designed as a safety net and which has been allowed to expand in an unsustainable way far beyond its original purpose, (2) welfare, which has been enormously counter-productive to its original intent (although this is one of the few programs which was largely rolled back in the mid-1990s), and (3) a steeply progressive individual income tax system

You might say that this is ancient history, and you don't hear much about Progressives today. However, keep in mind that one of the Democratic presidential candidates in 2008, Hillary Clinton, said in one of the Presidential debates that she would rather be called a "modern day progressive" than a liberal.[11] She obviously chose her words carefully, and she wanted to be associated with the views and the works of the Progressives. Unfortunately, it appears from this that Progressivism is alive, and this is why constant vigilance is required to limit the damage any resurgent Progressives might do.

The landscape of history is littered with examples of societies which started out with well-conceived ideas and ended up losing their civilizations because, among other things, they were not able to maintain fidelity to the original concepts through the generations. How can a people maintain commitment to the founding principles if they don't know what they are?

One indication of the decline in our civilization which may be taking place is the demotion of classes in civics to the bottom of the curriculum in most high schools and colleges in terms of importance. The decline in the quality of education in this country is discussed in Chapter 8.

Another indication is the lack of integration of many modern-day immigrants. Happily, we are historically and by definition a nation of immigrants, and they have been absorbed in and become part of our society. However, more recent immigrant groups, which are rapidly growing in size, are not given the background in the fundamental principles of the system of government which has made this country great, and, as a result, they don't have the same commitment to them.

Thus, whether we are talking about children who are being educated today or the stream of recent immigrants, on both of whom we have to rely for the preservation and further development of our way of life, we find that the knowledge of civics is very limited, if it exists at all. The result, based on history, is predictable and inevitable. George Santayana, the Spanish philosopher, said that "those who cannot remember the past are condemned to repeat it".[12]

Ignorance, indifference, and apathy with respect to the founding documents are commonplace today among the American people. Our system of government requires both knowledge and constant vigilance by every citizen to withstand the challenges with which every civilization is confronted. Taking it for granted, which I think has characterized the public attitudes over the last few decades, will guarantee that it will fail.

It is not too late to restore our civilization to greatness and preserve it for our posterity, but the time left to save it is growing short, as will be demonstrated in Chapter 2.

Conditions have deteriorated so much and we are so far from the provisions of the Constitution, that only a Revolution will restore Constitutional government. What is needed is not a military takeover. Nor is it a revolution of some small group of "elite" people with special interests or a revolution to make small changes in the institutions with which we live today. What is needed is nothing short of a Revolution for the American people to take back America and enable us to Restore the Future.

Bottom Line

- For good reasons, human beings decide to live in societies and figuratively sign a social contract.
- A broad range of forms of government has been tried, and most of them have failed to live up to the terms of the contract.
- The system of government designed by the Founding Fathers is the pinnacle of the development of systems of government to date.
- It has allowed this country to develop into the most successful country in the world and to provide a beacon for millions of people who understand the opportunity which our system of government can provide.
- However, the knowledge about and commitment to the basic principles of that system of government are declining rapidly.
- Without that knowledge and that commitment, fiercely defended against all internal and external challengers, it is inevitable that historians at some point in the future will write extensively about the "Decline of the American Republic".
- The clock is ticking, but it is not too late for aggressive action to, in Franklin's words, "keep the Republic" and Restore the Future.

CHAPTER 2
THE INDICTMENT AND CALL TO REVOLUTION

"...that whenever any form of Government becomes destructive of these ends, it is the Right of the People to alter or to abolish it, and to institute new Government, laying its foundation on such principles and organizing its powers in such form, as to them shall seem most likely to effect their Safety and Happiness."

Thomas Jefferson –
Declaration of Independence

Background

Recall that a significant part of the Declaration of Independence was devoted to a list of grievances which the colonies had experienced under King George, such as: (1) "He has erected a multitude of New Offices, and sent hither swarms of officers to harass our people, and eat out their substance", and (2) "... imposing Taxes on us without our consent". The list was prepared to set the background for worldwide approval of the call for revolution first suggested by John Locke under circumstances like those which existed when the Declaration was written.

It is interesting to note that these two specific grievances are no less significant and relevant today than they were in 1776 when they were first enunciated, and no less actionable.

There is a broad parallel today in the grievances of citizens of this country with modern government. If it can be shown that the natural rights of citizens are being trampled on, that the Federal government is not acting as it was intended to do under the Constitution, and that actions and laws are not examined for Constitutionality before they are taken or implemented, then a revolution of the people (non-military) against their elected representatives to return the Presidency and the Congress to a commitment to meeting the requirements of the founding documents is both necessary and proper. This will be the Second American Revolution, and its purpose is to Restore the Future.

As an example, according to the Heritage Foundation, a conservative research organization based in Washington, DC, a reporter asked Nancy Pelosi, the former Speaker of the House of Representatives, "Madam Speaker, where specifically does the Constitution grant Congress the authority to enact an individual health insurance mandate" (a part of Obamacare requires each person to purchase health insurance or pay a fine). Her response was "You can put this on the record. That is not a serious question".[1]

Heritage continues: "The Congressional Budget Office (CBO) disagrees. In 1994, the CBO said of an individual mandate to buy health insurance:

"A mandate requiring all individuals to purchase health insurance would be an unprecedented form of federal action. The government has never required people to buy any good or service as a condition of lawful residence in the United States. An individual mandate would have two features that, in combination, would make it unique. First, it would impose a duty on individuals as members of society. Second, it would require people to purchase a specific service that would be heavily regulated by the federal government...

If the individual mandate is Constitutional, then Congress could do anything".[2]

This is a very fundamental issue which goes to the heart of our constitutional government, and the Speaker feels that it is not a "serious question"? This is a naked abuse of power and the Constitution, and it should be totally unacceptable to the American people.

The First Revolution was fought to establish the extraordinary ideas of natural rights of American citizens, gain independence because those rights were being systematically violated, and create a lasting form of government to protect and maintain those rights.

As you will see from the information shown below, it is both critical and justified to carry out a non-violent Second Revolution in this country to Restore the Future and return our government to the specific role for which it was originally intended – protecting the natural rights of the citizens of this country. This is what the rise of the Tea Party in recent years is all about. A better name for the Tea Party would be the "Constitutional" party.

The Indictment

A partial list of specific grievances against government in the US at the current time is provided below. This list is presented in the same spirit as Jefferson's list in the Declaration of Independence and for the same reasons. Exposing the grievances is the first step to building support in the country and around the world for a revolution in the way our government is conducted. At the end of this list is an in-depth discussion of each of five other major grievances.

As Thomas Paine said in his "Crisis", 9/23/76:

These are the times that try men's souls. The summer soldier and the sunshine patriot will, in this crisis, shrink from the service of their country: but he that stands by it now, deserves the love and thanks of man and woman. Tyranny, like hell, is not easily conquered: yet we have this consolation with us, that the harder the conflict, the more glorious the triumph. What we obtain too cheap, we esteem too lightly: it is dearness only that gives every thing its value. Heaven knows how to put a proper price upon its goods; and it would be strange indeed if so celestial an article as freedom should not be highly rated.

Ronald Reagan said:

Freedom is never more than one generation away from extinction. We didn't pass it to our children in the bloodstream. It must be fought for, protected, and handed on for them to do the same.

What Thomas Paine said in 1776 and what Ronald Reagan said 200 years later once again sound the call to the Second American Revolution.

Specific Grievances

Let's consider the following specific departures from the founding documents as examples of what is happening today.

- There is a large scale effort to revise the Constitution through legislative action or judicial fiat without going through the process which the Founders designed for making amendments to the Constitution. The President of the United States is sympathetic to this movement.

    Listen to his words when he was in the Illinois Senate: "The Constitution, he said, had 'deep flaws'. He faulted the Supreme Court as well, because 'it didn't break free from the essential constraints that were placed by the Founding Fathers in the Constitution."[3]

    Let's not forget that the President takes an oath of office in which he commits to support the Constitution. Based on the views expressed above, which are the parts of the Constitution to which the oath does not apply?

- Legal scholars are arguing that interpretation of our laws should be conditioned by what is happening in other legal systems around the world. This is preposterous on its face. Our laws, by definition, can only be understood to be legitimate in terms of our own Constitution. Nothing else can or should be considered in evaluating their legitimacy.

- The present administration and its supporters are sympathetic to the ideas of "One World Government" and a "World Court". What these ideas mean quite specifically is that our constitutionally created government and our legal system would be subordinated to institutions whose members have not been elected, over which we have no influence, and whose philosophies are almost certainly not the same as ours. It is certainly in our best interest to work with other governments and legal systems without destroying the integrity of our own, but subordination is unthinkable.
  - Even if it did make sense, why would we think that institutions of this type would actually succeed? If, for example, the United Nations cannot be made to work effectively and our interests cannot be protected, why would these institutions be any different?
  - Would you really want American soldiers to be tried in the World Court for "war crimes"?
  - Would you want the conduct of our own environmental policy to be constrained by legally binding obligations to some "super" environmental group?
  - In general, would you want to be under the influence of non-American institutions which could require that we give up something that we felt was in the best interest of this country for a greater good as defined by them?

An example of this is the effort by the Obama administration to support an Arms Trade Treaty proposed by the United Nations that will deal with arms control, including firearms. It probably will contain such provisions as gun registration and bans on guns. If this is the way it turns out, what has happened to our rights under the Second Amendment?

- The Founders carefully outlined the functions and responsibilities of the three parts of government, so that there would be no overlap and each part would have checks and balances on the others. Today, the Judiciary has begun to create policy, as opposed to determining the constitutionality of laws, when such policy has not been subjected to state or federal legislative review. The making of policy by the Judiciary was specifically prohibited by the Founders, and it clearly represents an unconstitutional end run around the Congress, which has the responsibility to make policy and enact legislation.

- While there is appropriate concern about extending the national debt limit and the massive, unprecedented, and totally unjustified budget deficits, there is another problem which is equally important but which does not receive the same attention.

"The federal government's financial condition deteriorated rapidly last year, far beyond the $1.5 trillion in new debt taken on to finance the budget deficit... The government added $5.3 trillion in new financial obligations in 2010, largely for welfare programs such as Medicare and Social Security. That brings to a record $61.6 trillion the total of financial promises not paid for. The $61.6 trillion in unfunded obligations amounts to $534,000 per household."[4]

This is a tremendous financial burden for future generations, and yet nothing is being done to address it.

- The President has systematically engaged in class warfare so that he can try to raise taxes to cover the unprecedented spending for which he is responsible. Despite the facts that the top 10% of incomes pay 70% of the taxes, marginal tax rates are already high, and those with these incomes create most of the jobs, the President pretends to know what your fair share is (over $250,000) and wants to raise taxes even more on the "wealthy".

The President's attitude is illustrated by the following quotation from a press conference on 7/11/11: "...and I do not want, and I will not accept, a deal in which I am asked to do nothing, in fact, I'm able to keep hundreds of thousands of dollars in additional income **that I don't need**, while a parent out there who is struggling to send their kid to college suddenly finds that they've got a couple of thousand dollars less in grants or student loans."[5] (emphasis added)

This is not equality of opportunity, and this is not the government living up to one of its primary responsibilities, which is to promote liberty. This is an undisguised effort to redistribute income and promote equality of condition.

- The current administration is moving as quickly as it can to create a system of socialized medicine. This idea was summarily rejected when it was first proposed by the Clintons in the early 1990s, and the majority of American people don't support it today. Some questions must be raised:
  - Did the Founders ever contemplate that the federal government would be by law responsible for providing health care to all citizens? This is not one of the enumerated powers of the federal government.
  - Are those promoting this concept not aware of the complete failure of socialized medicine systems in the UK and Canada, so much so that Canadians come to the US for critical health care as often as they can?
  - The data in Figure 2-1, provided by the United Nations International Health Organization, are samples, among many, of the differences in health care among the three countries. Do they support the claim about the superiority of health care in England and Canada?

Figure 2-1

	Cancer Survivor after 5 Years	Hip Replacement for Seniors within 6 Months	MRI Scanners per Million People
US	65%	90%	71
England	46	15	14
Canada	42	43	18

  - Do they not know that as a result of socialized medicine, there are no longer any major drug companies in Canada? Not surprisingly, President Obama has said that we should not expect major improvements in the state-of-the art in medicine in his plan (because the incentives have been eliminated).

- Has any thought been given to what happens if health care companies and health care providers end up in India or China, because they can operate more freely in other parts of the world?
- With the exception of national defense, where the federal government over the years, until recently, has generally fulfilled its Constitutional responsibilities to successfully prosecute wars when necessary and to maintain a state-of- the-art military capability, both offensively and defensively, has the federal government actually operated to a high level of performance and efficiency any major activity? What about the Post Office, which is not competitive and operates deeply in the red? What about the waste, fraud, and ineffectiveness of Medicaid and Medicare, which also operate in the red. What about Amtrak, which operates in the red? The list goes on and on.

As has been proven over and over again in modern history, governments are typically not good at <u>operating</u> anything (in the extreme, think about the experience with central planning in Russia). Efforts to bring more and more of the economy under government control are doomed to failure, but the attempted implementation of such efforts will bring our economy to its knees.

- The Constitution carefully creates a "separation of powers' between the federal government and the states. A specified list of powers is given to the federal government, and all other powers reside with the states. A careful examination of the extent of the powers which it has assumed reveals that that government has trampled on the separation of powers doctrine by assuming far more powers than the Founders originally intended.
- The Founders were concerned that since they had given so much power to the states to control their own destinies within the framework of the Constitution that the states might try to engage in practices which would disrupt interstate commerce. Therefore, they created the "Commerce" clause, giving the power to the federal government to regulate interstate commerce. This clause has been expanded to cover

almost anything the federal government wants to do, because it defines almost anything as interstate commerce.

The best current example of this is Obamacare, which is under review for constitutionality by several federal appeals courts and which has ended up in the Supreme Court. The Obama administration is arguing that it has a right to regulate the health care industry through its constitutional authority to regulate interstate commerce. It says that consumers who don't buy health care are nevertheless involved in interstate commerce, and the federal government can force them to buy health care or pay a fine.

- Rights have been created where none exist in the Constitution. There is nothing in the Constitution about a woman's right to choose to have an abortion or not. As a result, many legal scholars, including the Chief Justice of the Supreme Court, believe that Roe vs. Wade was wrongly decided. Independent of your view about abortion, abortion as a legal matter is an issue for the states to decide.
- The First Amendment to the Constitution says, "Congress shall make no law respecting an establishment of religion, or prohibiting the free exercise thereof...." The Founders, having had intimate experience with the state-supported Church of England, wanted to make sure that no state religion was established. However, they specifically established the free exercise of religion as a legitimate activity.

The modern idea of separation of church and State was developed out of thin air by Justice Hugo Black, based on a letter to a congregation in Danbury, Connecticut written by Thomas Jefferson. This subject will be discussed at length in Chapter 3, but Black's misinterpretation of the intent of the Founders has led to a significant decline in organized religion and increased secularism.

- The Founders were very concerned about "Tyranny of the Majority", and they designed systems of government specifically to keep it from developing. However, this whole idea has now been turned on its head. Now we are constantly being subjected to "Tyranny of the Minority". Anyone who disagrees with something and feels discriminated against is permitted to disrupt the whole society and change any of the societal conventions the majority supports, no matter how big the majority.

- In an overall sense, we have arrived at a point at which respect for the major institutions which have always been pillars of our society has declined to new lows. People today have little interest in or respect for: (1) government itself at any level (Congress has an approval rating below 20%), (2) education – the performance of American children is at best average among a group of major countries on standardized tests – education is addressed in Chapter 8, (3) churches and synagogues – interest is strong and growing only in mosques, (4) the institution of marriage as being between a man and a woman, and (5) self-reliance, as opposed to the demand for entitlements from government. The list goes on and on.
- An entitlement mentality has become entrenched in our national psyche. The impact of this is dramatically demonstrated in the charts below. I do not believe that there is anything in the Constitution which supports the level of spending on entitlements which will be discussed later. The pernicious thing about entitlement programs is that once they are started they grow apace with no natural limits. As Thomas Sowell, an economist, said, "If we become a people who are willing to give up our money and our freedom in exchange for rhetoric and promises, then nothing can save us."[6]

There is, however, one shining example of the rollback of a major entitlement program in the last twenty years, and that is welfare. The structure of and the incentives for welfare were dramatically changed in the 1990s, and the result was very positive – welfare rolls declined dramatically, and millions of citizens were transformed into productive workers. Welfare, as we knew it then, hardly exists today. After all, "The Constitution was never intended to guarantee people equal results in life, only guaranteed them equal opportunities."[7]

- We have now reached the point at which 53% of the people pay 100% of the taxes, according to the Tax Policy Center, a Washington research organization. Half the wage earners in this country pay taxes so that the other half doesn't have any taxes at all. Does this seem fair? Are those who do not pay any taxes at all "paying their fair share"? What is happening with the other 47%? As is the case in many western countries with similar situations, the other 47% is surviving on entitlements (see below).

    The Tax Policy Center also points out that the top 10% of earners pay about 70% of the total taxes collected. This is about the most progressive tax system which can be imagined. Is this what the Founding Fathers had in mind?

- The EPA has now issued a ruling that "greenhouse gases are dangerous pollutants", which allows it to regulate the emission of such gases, including carbon, across the entire economy. "With this doomsday machine, Mr. Obama hopes to accomplish what persuasion and debate among his own party manifestly cannot...The White House has opened a Pandora's box that will be difficult to close, that is breathtakingly undemocratic, and that the country... will live to regret"[8]. This is clearly an end run around the normal legislative process.
- Obama has created an indeterminate number of "czars" whose primary responsibility is to help him manage all governmental activities and the economy. What is remarkable about the creation of the czars is they are clearly not elected, and, what's worse, their appointments, in many cases, are not approved by Congress.

    The "Pay" czar, for example, has decided to limit the annual compensation of individual employees in all businesses in which the government has a stake to $500,000. While this is a very large amount of money, there is no Constitutional justification for the existence of a czar in the first place. Even if there were, there are two fundamental problems with allowing the czar to make a decision like this: (1) there is no way for anyone to know what the "proper" amount should be, even if you wanted to have one – the decision is totally arbitrary, and (2) this arbitrary interference in the economy will be counterproductive for economic growth, as was proved in the Soviet Union when it tried and failed miserably to make command and control decisions for the entire Russian economy.

Five Additional Major Grievances

In addition to the list above, there are five major additional grievances which have to be discussed in some depth: (1) the growth in public sector employment and compensation, (2) the growth of entitlement spending, (3) inadequate spending on national defense, (4) the unimaginable and perhaps fatal mismanagement of the country's finances, and (5) rampant hypocrisy in government.

Public Sector Employment and Compensation

USA Today has conducted a thorough study of the pay of federal workers during the recent recession (12/07-6/09), and here are some of the conclusions: (1) the number of federal employees making $100,000 per year increased by 46% during this period, (2) the number making $150,000 went up by 119%, and (3) the number making $170,000 or more increased by 93%.[9] This is in an environment in which the private sector lost 7.3 million jobs. This dramatic escalation in compensation seems to have been accomplished without any comment from the Pay czar.

While the figures for the last two years are particularly dramatic, they are really part of a long-term trend described in Figure 2-2. The data are provided in an article published by the Cato Institute[10] The average compensation (wages plus benefits) for federal workers has been growing at an annualized rate of 5.8%, while average compensation for workers in private industry has been growing at only 3.4%. As of the end of 2008, the average federal worker had a total compensation of $119,982, which was almost twice the total compensation of $59,909 for the average worker in the private sector.

Figure 2-2

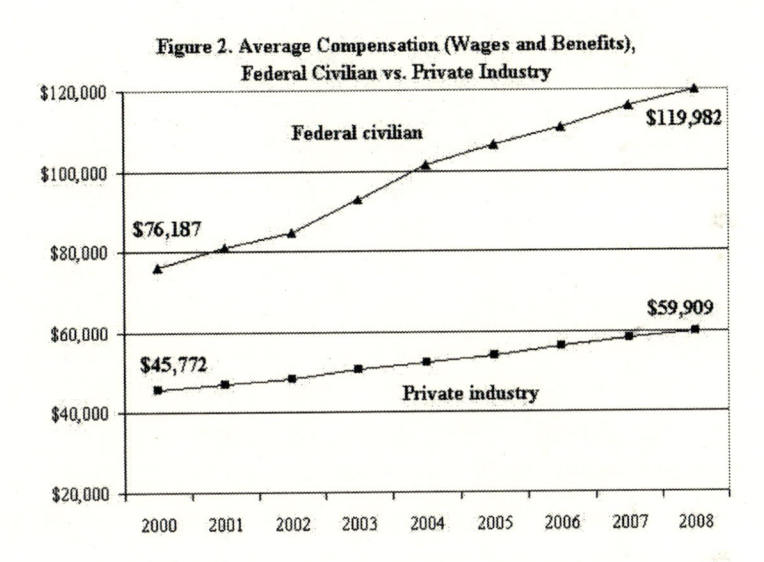

Figure 2. Average Compensation (Wages and Benefits), Federal Civilian vs. Private Industry

An update to this outrageous picture was provided by USA Today in a front-page article on 8/9/10. "Federal workers have been awarded bigger average pay and benefit increases than private employees for nine years in a row. The compensation gap between federal and private workers has doubled in the past decade.

Federal civil servants earned average pay and benefits of $123,049 in 2009, while private workers made $61,051 in total compensation, according to the Bureau of Economic Analysis. Federal civil servants make more than twice as much as their counterparts in private industry. The Federal compensation advantage has grown from $30,145 in 2000 to $61,998 last year."

The data for different classes of workers in 2009, according to the USA Today article, are shown in Figure 2-3:

Figure 2-3

	Salary	Benefits	Total
• Federal civilian	$81,258	$41,791	$123,049
• State and local	53,056	16,587	69,913
• Private	50,462	10,589	61,051

While state and local employees make more than their counterparts in private industry, the really dramatic and inexplicable difference is between the compensation for federal civilian (excluding the military) employees compared to their private sector counterparts. Furthermore, the comparison continues to get worse.

It is interesting to note that on 1/22/10, the US Bureau of Labor Statistics (BLS) in its annual report on union membership reported that: (1) for the first time, the majority of union workers are government workers (7.9 million) rather than private sector employees (7.4 million), and (2) 7.2% of the private sector workers were unionized, whereas union membership increased to 37.4% of government workers. In California, according to the New York Times on 1/24/09, the percentage is 85%.

The Wall Street Journal reports that, "According to the BLS, in 2009, the average state or local public employee received $39.66 in total compensation per hour versus $27.42 for private workers."[11] This is a premium of about 45%.

This is an extraordinary state of affairs when you consider that historically public employees were willing to accept less than their counterparts in private industry in total compensation, because their job security is generally so much higher than is the case in private industry. In effect, the public sector employees are "double dipping". They have higher job security and much higher total compensation.

Let's go behind the numbers to see how devastating for state and local governments this compensation premium has become. "In 2008, almost half of all state and local government expenditures, or an estimated $1.1 trillion, went toward the pay and benefits of public workers."[12] A recent study by the Bureau of Labor Statistics provides the following information about the sources of the 45% premium indicated above.[13]

Public employees earn salaries which are about 1/3 higher than the salaries of the employees in private industry. However, the difference in benefits for the two groups is astounding. Public sector employees at the state level have benefits which are 60% higher than private employers offer. Finally, public employee health benefits are more than 100% greater than what is offered in private industry.

"By the way, nearly this entire benefits gap is accounted for by *unionized* public employees. Nonunion public employees are paid roughly what private workers receive."[14] These compensation differentials demonstrate how powerful public employee unions have become and point the way to the major part of the solution to exploding state and local budget deficits.

However, this is not all. According to an article on the front page of USA Today on 9/29/11, "Retirement programs for **former** [emphasis added] federal workers – civilian and military – are growing so fast they now face a multi-trillion dollar shortfall nearly as big as Social Security's...The retirement programs now have a $5.7 trillion unfunded liability, compared with $6.5 trillion shortfall for Social Security. An unfunded liability is the difference between a program's projected costs and its projected revenue..."

There is a similar enormous financial problem for state and local governments which receives very little press but which is going to create extraordinary pressures on state and local government budgets for years to come even if they solve their short-term deficit problems.

The BLS study cited above shows that the value of defined benefit pension plans included in the total compensation figures for state and local employees is 700% of the value of these plans in private industry. For years, these employees have been granted ever larger pension and other long-term benefit payments, without anyone really noticing. The result is devastating.

"A 2009 study by two economists published in the Journal of Economic Perspectives estimated that these government pensions are underfunded by $3.2 trillion or $27,000 for every American household."[15] Underfunding means that state and local politicians have promised pension payments which are $3.2 trillion more than the funds which exist to pay those benefits. Clearly, only two things can happen: (1) these benefits will have to be slashed, or (2) state and local taxpayers will have to make up the difference out of their current earnings.

There is no justification whatsoever for taxpayers to be assessed what is estimated above to be $27,000 per family to make up for the extraordinary benefits which politicians have provided to public sector employees over the years.

"Representing government employees has changed the union movement's priorities. Unions now campaign for high taxes on Americans to fund more government spending. Congress should resist government employees unions' self-interested calls to raise taxes on workers in the private sector."[16]

The following quotation helps explain what is going on with public sector unions and why they want the increase in government spending. "Public employee unions have a lucrative racket: They essentially leverage the tax dollars they receive in dues from the salaries and benefits of their members to lobby for more tax dollars to secure even fatter pensions and pay"[17]

This has happened because politicians have been unable to resist the temptation to: (1) grow government without constraint (more government employees), and (2) increase salaries and benefits without the constraints which exist in the private sector. The private sector, after all, has profit and service requirements; the public sector has no profit motive, and it therefore is not motivated to operate efficiently.

It is not at all clear why there should be such a substantial premium for compensation for public sector employees compared to those in the private sector. In fact, it is not clear why there should be any premium at all. In any case, the high percentage of unionization and the dramatic compensation premium have created considerable pressure on local and national budgets.

Government spending at all levels is out of control and must be reduced. Making reductions in bloated public sector employment, which is a significant portion of government spending, at least at the local level, given the information above, will be very difficult. However, doing so is the sine qua non for restoring sanity to government budgets at both the federal and state levels.

It is extremely helpful to see that the governors of both Virginia and New Jersey, who were elected several years ago, have said that "enough is enough". They have refused to raise taxes to cover large budget deficits, and they are beginning to attack the level and causes of government spending. Republican governors all over the country are doing the same things.

Entitlements

Figures 2-4 through 2-9 are charts prepared by the Heritage Foundation[18], and they are based on projections by the Congressional Budget Office and the White House Office of Management and Budget.

Figure 2-4 shows the extraordinary growth in entitlement spending.

Figure 2-4

Entitlements Will Consume All Tax Revenues by 2049

If the average historical level of tax revenue is extended, spending on Medicare, Medicaid and the Obamacare subsidy program, and Social Security will consume all revenues by 2049. Because entitlement spending is funded on autopilot, no revenue will be left to pay for other government spending, including constitutional functions such as defense.

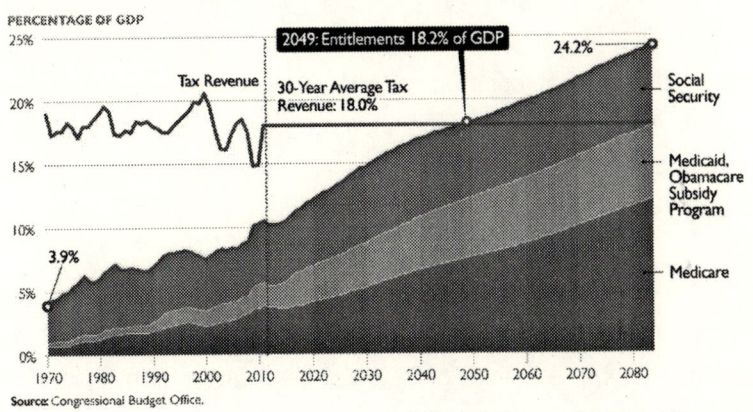

Source: Congressional Budget Office.

Entitlements Chart 1 • 2011 Budget Chart Book • heritage.org

There are five important points to note about this Figure:

- For the last 40 years, the annual total tax revenue of the federal government has fallen in a range of about 15-21% of Gross Domestic Product (GDP), and the average for that period has been 18.0%.
- As of 2010, the three entitlement programs shown – Medicaid, Medicare, and Social Security - accounted for about 10.0% of total GDP, and they collectively were growing at a very high rate.
- By the year 2049, these three entitlements alone will account for 18.2% of GDP, which is more than the average historical amount of total federal tax revenue as a percentage of GDP.
- What this means is that just paying for entitlements in 2049 will require the federal government to collect as much revenue as it has collected on average for the last 40 years for the operation of the entire government. There would not be room for any other spending. The only alternative would be to increase taxes significantly.
- The unrestrained growth in these programs would account for 24.2% of GDP in 2085, and this would require an additional massive increase in taxes

Maybe there are alternatives to this cataclysmic result. There are three primary ways the Federal government can come up with funds for its spending; (1) taxes (see below), (2) money creation, or (3) borrowing. Massive money creation as a long-term approach is clearly inflationary. Borrowing, if anyone would be willing to lend under the circumstances described above, simply postpones the need to come up with the money, creates enormous debt service requirements, and burdens future generations with such a mountain of debt repayment obligations that the country would almost certainly go bankrupt.

I think it is problematic, in any case, to expect that the rest of the world will be willing to provide an unlimited amount of financing for the government to operate in this profligate way, and, therefore, most of the funds, by default, would have to come from increased taxes and money creation.

These long-term projections don't reflect the latest information about the Social Security program. "... unless we get an immediate and sharp recovery, the revenues of the [Social Security] trust fund will be tracking lower [than expected] for a number of years."[19] Furthermore, according to an article in the New York Times published on 3/25/10, "This year, the [Social Security] system will pay out more in benefits than it receives in payroll taxes, an important threshold it was not expected to cross until at least 2016, according to the Congressional Budget Office."[20]

Clearly, this is a time bomb waiting to explode. Under the assumptions outlined above, as shown in Figure 2-4, tax rates would have to more than double, as shown in Figure 2-5, according to the Congressional Budget Office, unless there were cutbacks in other areas, which is unlikely. Instead of recognizing this problem and beginning to deal with it, politicians today only focus on the short-term and on programs which will make the situation worse, like Obamacare.

Figure 2-5

Hiking Taxes to Pay for Entitlements Would Require Doubling Tax Rates

The cost of Medicare, Medicaid, and Social Security is rising substantially. Paying for this spending solely through federal income tax increases would require more than a twofold increase of current tax rates, even for the lowest tax bracket.

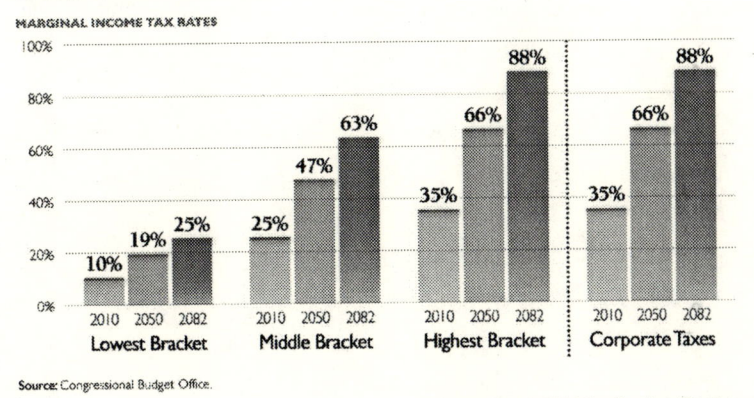

MARGINAL INCOME TAX RATES

Bracket	2010	2050	2082
Lowest Bracket	10%	19%	25%
Middle Bracket	25%	47%	63%
Highest Bracket	35%	66%	88%
Corporate Taxes	35%	66%	88%

Source: Congressional Budget Office.

Entitlements Chart 6 • 2011 Budget Chart Book • heritage.org

This is a preposterous outcome which will never happen. However, it illustrates the magnitude of the problem and the desperate need to control entitlement spending.

I show this chart at this time to demonstrate how far we have come from the limited government of enumerated powers outlined in the Constitution. Do you think that the Founders had anything like this in mind? Is this the government whose powers are limited as enumerated in the Constitution? This is an outrage, and voters in this country must quickly come to their senses and put a stop to what is happening, or they will have "lost" their country.

It is easy to see at a glance how devastating the growth of entitlement spending and the necessary federal government revenue generation would be, assuming everything else is equal. However, everything else is not equal in this situation.

To obtain the necessary tax revenue, even if it were possible, the money would have to come from the private sector. What this means is that every dollar of increased taxes is taken from those parts of the economy which produce growth and create jobs and is in turn invested in the public sector, which is known for extraordinary inefficiency and fraud. Imagine how costly this substitution is to economic growth.

Not surprisingly, the evidence is overwhelming in study after study that increases in spending by the Federal government reduce overall economic growth. Comments from one of them are shown below:

"Most government spending has historically reduced productivity and long-term economic growth due to:

- **Taxes**. Most government spending is financed by taxes, and high tax rates reduce incentives to work, save, and invest – resulting in a less motivated workforce as well as less business investment in new capital and technology. Few government expenditures raise productivity enough to offset the productivity lost due to taxes;
- **Incentives**. Social spending often reduces incentives for productivity by subsidizing leisure and unemployment. Combined with taxes, it is clear that taxing Peter to subsidize Paul reduces both of their incentives to be productive, since productivity no longer determines one's income;

- **Displacement**. Every dollar spent by politicians means one dollar less to be allocated based on market forces within the more productive private sector. For example, rather than allowing the market to allocate investments, politicians seize that money and earmark it for favored organizations with little regard for improvements to economic efficiency; and,
- **Inefficiencies**. Government provision of housing, education, and postal operations are often much less efficient than the private sector. Government also distorts existing health care and education markets by promoting third-party payers, resulting in over-consumption and insensitivity to prices and outcomes. Another example of inefficiency is when politicians earmark highway money for wasteful pork projects rather than expanding highway capacity where it is most needed."[21]

The latest study about the massive stimulus program passed in early 2009 confirms what all the other studies have shown and adds additional perspective. The study shows that: "(1) [there] is no statistical correlation between unemployment and how the $862 billion was spent; (2) Democratic districts received one-and-one-half times as many awards as Republican ones; and (3) an average cost of $286,000 was awarded per job created."[22]

National Defense

One of the most basic responsibilities of government in any society is national defense, which is discussed in depth in Chapter 7. Many people have the impression that defense spending accounts for a significant percentage of our gross domestic product (perhaps 10-20%). If we were in fact spending as much as people imagine, the Federal government might be living up to its responsibilities.

The actual situation is shown in Figure 2-6:

Figure 2-6

Obama's Budget Would Reduce National Defense Spending

Adequate funding for the core defense program is crucial for the military to fulfill its constitutional duty to provide for the common defense. Yet defense spending has fallen below its 45-year historical average despite ongoing operations in Iraq and Afghanistan.

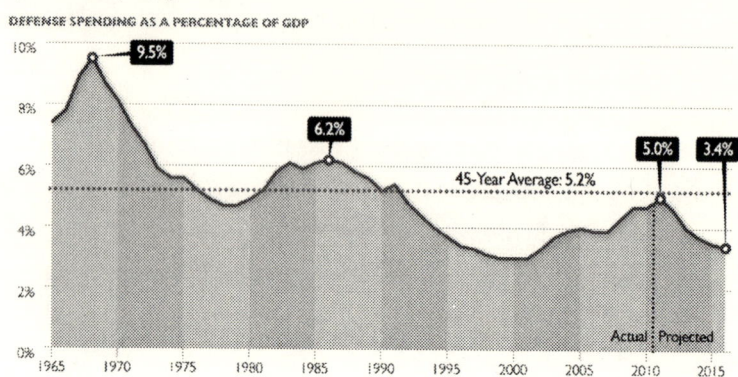

DEFENSE SPENDING AS A PERCENTAGE OF GDP

- 9.5%
- 6.2%
- 45-Year Average: 5.2%
- 5.0%
- 3.4%
- Actual | Projected

Source: White House Office of Management and Budget.

Federal Spending Chart 8 • 2011 Budget Chart Book • heritage.org

National defense spending has historically averaged 5.2% of GDP. The Vietnam buildup, the Cold War buildup under Ronald Reagan, and the sharp reduction in defense spending during the Clinton years are clearly visible. I would argue that even the historical average level of spending in peacetime is grossly insufficient for the world in which we currently live.

However, it is striking that Obama's budget would reduce national defense spending to pre-9/11 levels by 2016, which is right around the corner, despite the on-going war in Afghanistan This does not include the additional defense cuts which the administration would like to make as part of the deal to increase the national debt limit.

It is not clear what the right percentage is, but this is almost certainly not enough, at least under the present circumstances of terror directed at the US. If overall defense spending is reduced and if war spending is maintained, non-war related defense spending will have an even greater percentage reduction than defense spending as a whole.

What has happened is that the defense budget, as indicated above, has become a residual of spending for other things in the budget. A very high percentage of the national budget is spent on entitlement programs which are generally on auto-pilot and not subject to effective management, and defense is squeezed out.

This is just the reverse of the way that government should operate. The first principle should be to determine how much of the national budget should be spent on keeping us safe, and therefore other spending categories would be the residual. If we are not spending enough to protect ourselves, the rest of the budget doesn't matter anyway.

Does anyone really think that the present level of spending meets the Constitutional requirement for the federal government to protect and defend the country?

Fiscal Mismanagement

The uncontrolled amount of entitlement spending, even with the drawdown in defense spending needed to provide for our security, is driving federal spending, budget deficits, and borrowing to unprecedented levels.

Figure 2-7

Runaway Spending, Not Inadequate Tax Revenue, Is Responsible for Future Deficits

The main driver behind long-term deficits is government spending—not low revenues. While revenue will surpass its historical average of 18.0 percent of GDP by 2021, spending will shoot past its historical average of 20.3 percent, reaching 26.4 percent in the same year

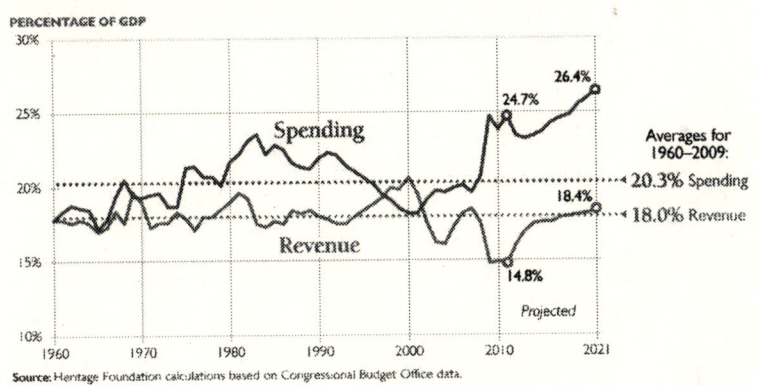

PERCENTAGE OF GDP

Averages for 1960–2009:
20.3% Spending
18.0% Revenue

Spending: 26.4%, 24.7%, 18.4%
Revenue: 14.8%
Projected

Source: Heritage Foundation calculations based on Congressional Budget Office data.

Federal Spending Chart 11 • 2011 Budget Chart Book • heritage.org

Let's look first at spending, as illustrated in Figure 2-7.

Almost every year since 1962, federal spending has exceeded federal revenue, with the exception of a brief period at the beginning of this decade. However, as recently as 2007, the gap was not unprecedented. Since then, the difference between spending and revenue is truly astonishing. Revenues have turned down, and spending has gone ballistic. Based on estimates for 2011, spending will exceed revenue by 10 percentage points of GDP, and spending is greater than revenue by almost 2/3.

Since 1960, spending has averaged 20.3% of GDP, and revenue has averaged 18.0%. By 2021, revenue as a percentage of GDP is expected to be at a level of 18.4%, which is slightly above the long-term average, but spending is expected to be even higher than it is now – 26.4% versus 24.7% - and at that time, spending will be 8 percentage points of GDP higher than income.

These figures provide a dramatic demonstration that what we have in this country is a spending problem and not a revenue problem. It is clear that raising tax rates and trying to collect more revenue is an indefensible strategy compared to what is normal. It is equally clear that spending is out of control and has to be slashed.

Figure 2-8

Federal Budget Deficits Will Reach Levels Never Seen Before in the U.S.

Recent budget deficits have reached unprecedented levels, but the future will be much worse. Unless entitlements are reformed, spending on Medicare, Medicaid, and Social Security will drive deficits to unmanageable levels.

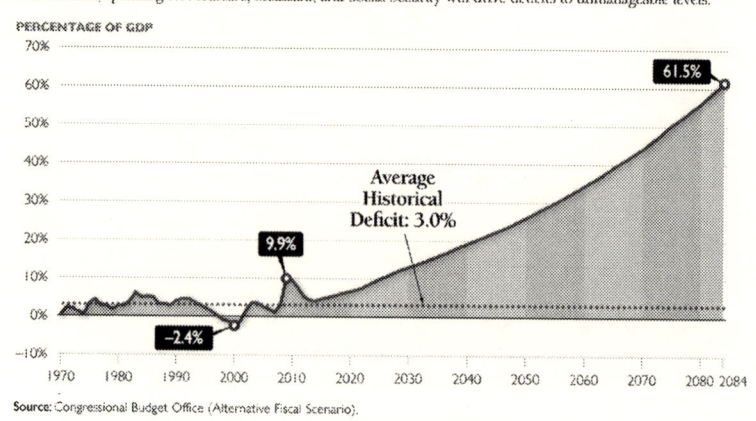

Source: Congressional Budget Office (Alternative Fiscal Scenario).

Debt and Deficits Chart 6 • 2011 Budget Chart Book • heritage.org

As shown in Figure 2-8, this level of spending would push Federal deficits to levels never before seen in this country. Even if the 2001 and 2003 tax cuts are allowed to expire and the AMT is not fixed, the Federal budget deficit in 2082 would be 61.5% of GDP, compared to the long-term average of 3.0%, or more than twenty times larger. It is difficult to relate to something which is this stunning. The current level of 9.9% of GDP is itself more than three times the historical average

Finally, the level of spending described in Figure 2-7, which is expected to continue, would send outstanding government debt to levels not seen since the sharp buildup in debt to finance the Second World War, as shown in Figure 2-9:

Figure 2-9

Obama's Budget Would Send Federal Debt to Levels Not Seen Since World War II

In 2008, publicly held debt as a percentage of the economy (GDP) was 40.3 percent, nearly four points below the postwar average. Since then, the debt has increased more than 50 percent, and the President's FY 2012 budget would more than double it to 87.4 percent by 2021.

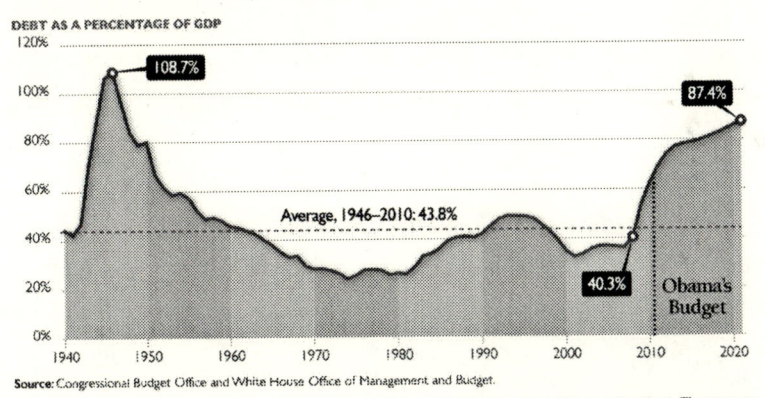

DEBT AS A PERCENTAGE OF GDP

- 108.7%
- Average, 1946–2010: 43.8%
- 40.3%
- 87.4%
- Obama's Budget

Source: Congressional Budget Office and White House Office of Management and Budget.

Debt and Deficits Chart 3 • 2011 Budget Chart Book • heritage.org

The long-term average for debt as a percentage of GDP has been 43.8%. At the end of 2008, debt was at 40.3% of GDP. Under Obama's budget, debt as a percentage of GDP would more than double to 87.4% by 2020. 2020 is obviously less than ten years away. Such a level should be summarily rejected out-of-hand by every American, regardless of political affiliation.

Figures 2-1 through 2-9, which show, among other things, rapid growth in the compensation of federal employees, an unbelievable growth in entitlement spending, a potentially dangerous reduction in defense spending, and mind-boggling growth in overall government spending and related debt, describe a government which is out of control and completely out of touch with the American people.

Few people would believe that these situations result from a government which is reflecting what the people want and what they think is in their interest, as opposed to reflecting a fundamentally dangerous ideology, political greed, back-room tradeoffs, and personal arrogance.

As you reviewed these Figures, it must have occurred to you that, in the absence of major policy changes, the chickens would come home to roost in serious and unpredictable ways. A warning of what could happen was issued on 8/5/11, when Standard & Poors, one of the major credit rating agencies, downgraded the rating on US debt to AA+ from AAA. This is the first downgrade of US debt in the history of the country.

There are 16 countries which now have a AAA credit rating, including France, Germany, the United Kingdom, Canada, Australia, Singapore, and Hong Kong. The only other country with a AA+ rating is Belgium.

Once a country is downgraded, it is difficult to get the rating back, and doing so can take a long time. As an indication of this situation, S&P goes beyond this downgrade. "The outlook on the long-term rating is negative. We could lower the long-term rating to AA within the next two years..."

This embarrassing situation is directly the result of the Obama Administration's extraordinary and unprecedented growth in spending, its lack of leadership in addressing the massive unfunded liabilities for Social Security, Medicare, and Medicaid, as outlined above, the inadequacy of the recent debt limit extension deal in cutting spending, and what S&P calls the "weakening in the effectiveness, stability, and predictability of American policymaking and political institutions".

The President cannot escape ultimate responsibility for this unfortunate development. As President Harry Truman famously said, "The buck stops here".

Hypocrisy

One of the reasons why so many people have little respect for and a very low opinion of the government officials which they elected is that, both individually and collectively, they display an arrogant hypocrisy.

Examples abound:

- They draft a hugely expensive, health care rationing and, in my view, unconstitutional health care system which affects 1/6th of the national economy. Yet they refuse to apply this massive legislation which few have read and fewer really understand to themselves. Does that make sense? If it such a good system, why aren't they eager to join? Why don't they demonstrate their conviction that this is somehow a better system than we have now by volunteering to be included?
- Congress and other Federal officials have a well-designed, effective, and heavily used investment program for retirement which they refuse to provide for the people they represent. As a result, people are flailing around trying to figure out what to do. Does this sound as if your elected officials are serving their constituents, or does it appear that they have created special privileges for themselves?
- The President pledges to eliminate earmarks, yet major legislation in the last year continues to include earmarks. Who benefits from spending taxpayer money on special projects which don't always see the light of day and which in many cases are designed to help politicians stay in office?
- The President pledges to have transparency about and time to review legislation before it comes up for a vote. For the first two years of his presidency, an arrogant Congress refused to do either, to the point where the bills were so long and so complicated that the Congress itself could not even read the bills before they are voted on. Does this sound

like representative democracy, or does it sound like public officials acting to implement their own agendas which they have no interest in sharing with the American people?
- The Obama administration promises honesty in government. Yet many backroom deals with certain senators and representatives are made which cost hundreds of millions of taxpayer dollars and which are not disclosed for the purpose of passing legislation which the taxpayers have indicated they do not want. Are these representatives and senators representing their constituencies, as they should in a representative government, or are they representing only themselves?
- Perhaps the clearest example of hypocrisy in government is the voucher program in the District of Columbia which provides for a small number of public school children the opportunity to go to private schools in the area. Public schools in the District are among the worst in the country.

The program has been a big success as judged by: (1) the performance of the children who have gone to private school compared to those who have not, (2) the fact that parents know this and are voting with their feet to get their kids into this program, and (3) the fact that there is a huge demand for a very limited number of participants. However, Congress in its infinite wisdom has: (1) set an arbitrary limit for the number of participants, (2) refused to add to the number of current participants, and (3) decided only recently to maintain the existing program, after earlier deciding that it should be terminated.

The hypocrisy of the Congress in education is shown in Figure 2-10[23]:

Figure 2-10

EDUCATION

Congress keeps school choice to itself

Four of every 10 members of Congress have sent at least one child to private school, and one in five went to private school themselves. The lawmakers, however, continue to reject choices outside public schools for less-affluent Americans.

Percentages that sent a child to private school or kept kids in public school:

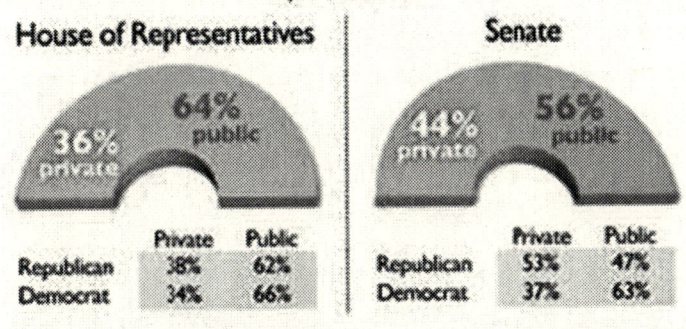

House of Representatives

36% private / 64% public

	Private	Public
Republican	38%	62%
Democrat	34%	66%

Senate

44% private / 56% public

	Private	Public
Republican	53%	47%
Democrat	37%	63%

For research on school choice, go to heritage.org and click on Education.

Sources: Heritage Foundation survey conducted Feb. 13 to March 13, 2009; previous surveys. The Heritage Foundation

---

Four of every ten members of Congress have sent at least one child to private school, and one in five went to private school themselves. Even President Obama chose, without much fanfare or criticism, to send his kids to private school. When these politicians make this kind of a choice, they are doing it for a good reason – they think their kids will get a better education. Why shouldn't other parents have the same choice? What monumental arrogance it is to deny them that opportunity.

"A Mind is a Terrible Thing to Waste" is the slogan of the United Negro College Fund. Minds are being wasted at a great rate in the Washington, DC public schools. How ironic it is that most of them are African-American.

Economic Freedom

To give some perspective on what these abuses have produced, I would like to cite the 2011 Index of Economic Freedom published by the Wall Street Journal and the Heritage Foundation. The Index is a comprehensive review of 183 countries based on information in ten separate areas – five focus on the regulatory burdens government places on business, three consider the size and intrusiveness of government, and two focus on property rights and freedom from corruption – to determine to what degree a country is economically free.

Why is this index important? "Detailed analyses have found that citizens in countries with the highest scores on the Index enjoy much higher standards of living than their neighbors in countries that are less free. Higher levels of economic freedom also go hand-in-hand with broader indications of both economic and social well-being."[24]

Where does the US stand? The 183 countries are grouped into five categories from the most to the least free, and the top group is called "free". In 2009, the US dropped out of the "free" category, in which it has been for all 16 years the Index has been published, into the "mostly free" category. While the US is still in the mostly free category, it's numerical rank has continued to fall to #9, behind such perennial leaders as Hong Kong and Singapore, and the trend is not encouraging. If this trend continues, the US will gradually become less free, and its relative standing in the world will continue to decline.

Conclusion

Reflecting on this list of grievances should be very unnerving. The list is so comprehensive and so endless that it makes you despair for the future of the country.

All of the information in this chapter is evidence that the politicians are working for themselves and counting on the citizens to acquiesce in the most monstrous government behavior which is imaginable in a representative democracy. They have the arrogance to assume that we are working for them and not the other way around. There is no other plausible explanation for the abuses which have been discussed in this chapter.

The recourse in a representative democracy is to do what the Founders did – overthrow the system of tyranny and those responsible for it. We must replace many of our representatives and reorient the others so that we get the responsible government which was created by the Founders and which we deserve.

It is never enough to present an indictment like the one presented in this book or, for that matter, any other criticism without proposing a well-considered-and well-reasoned proposal to remedy the situation. The Conclusion, which is Chapter 10, provides the road map which is required to Restore the Future.

Bottom Line

- There is a long list of specific grievances with respect to the way in which modern government functions, any one of which could be discussed at some length.
- I decided to do that with respect to five of them: (1) the growth in public sector employment and compensation, (2) the growth of entitlement spending, (3) inadequate spending on national defense, (4) the unimaginable and perhaps fatal mismanagement of the country's finances, and (5) the rampant hypocrisy in government.
- These grievances clearly demonstrate that our modern government is not functioning in the way it was designed to do by the Constitution, and that it has so far exceeded its constitutional responsibilities that we are in danger of losing our Republic.
- Under these circumstances, when the government is compromising the protection of the natural rights so brilliantly described in the Declaration of Independence, as the Declaration itself says, it is time to create a Second (non-violent) Revolution to force our government and our elected representatives to abide by the Constitution.

- Happily, the very same document has given us the power to effect the necessary changes by replacing our current representatives with representatives who are committed to restoring constitutional government and putting us on a path on which we can preserve the country we created for at least another 220 years.

CHAPTER 3
MORALITY AND RELIGION

"Can the liberties of a nation be thought secure when we have removed their only firm basis, a conviction in the minds of the people that these liberties are of the gift of God?"

Thomas Jefferson

Chapter 2 presents the indictment of the way our country is being run and the massive departures from Constitutional principles which are taking place in considerable detail. However, it doesn't address the root causes of the potential tailspin in which we find ourselves as a result. This chapter begins the process of identifying and appraising the status of these root causes. Doing so will provide a clear basis for determining what we need to do to Restore the Future.

In general, I have always thought that it is not enough to criticize what is going on with respect to any issue, however eloquent the criticism. That criticism must be accompanied by the identification of specific actions which have a good chance of rectifying the situation, or the criticism is not valid. I attempt to meet this burden of supporting the indictment in this and subsequent chapters.

Right and Wrong

One of the most fundamental debates in philosophy has always been between "moral absolutists" and "cultural relativists". The moral absolutist believes that there are certain aspects of societies which are "right" in a moral sense, independent of time and place. The cultural relativist believes that what is "right" in a moral sense is completely dependent on the time and place in which you find yourself.

Why is this debate important? From a social standpoint, if the people who live in a society do not know what is wrong, then everything by definition is right. Yet, the society must have some concept of right and wrong for the society to exist in the first place. People don't voluntarily become part of a society in whose moral concepts they don't believe or one in which what is right today is wrong tomorrow. Without some sense of morality, a society would fall apart. The question then is not "Can a society exist without a concept of morality?". Rather, the question is, "How is what is moral determined?."

It is easy to be a cultural relativist. Everyone knows that people change, institutions change, and societies change, and it seems unrealistic to assume that primitive societies and modern societies have anything in common from a moral standpoint. You just throw up your hands and say "it depends".

On the other hand, it is much more difficult intellectually to be a moral absolutist. The challenge is to identify those aspects of society which you would know would be right and wrong if you parachuted into any society with no background of it, coming from a different time and place. I would like to spend some time defending the moral absolutist, since the distinction between the two kinds of views about what is moral is so critical to the survival of our country.

What about truth-telling? It is hard to envision any society in which truth-telling is not a virtue. People in a society have to believe that most of the members of the society tell the truth most of the time. There are two other possibilities: (1) people have such different concepts of what the truth is that no society is possible, or (2) everyone always lies, in which case this becomes the new "truth". It seems clear to me that truth-telling is a prerequisite for society to exist and, therefore, an absolute standard for determining right and wrong.

The same general discussion applies to stealing. Is it possible to contemplate a society in which people steal from each other with impunity? The leaders of the society may take what they can get from the people, but is it plausible that ordinary members of society would tolerate stealing from each other? Isn't this a moral absolute?

What about not contributing to the welfare of the society as a whole – honoring a chief, paying allegiance to a king, and participating in the common defense? The point here is not that the allegiance and honoring are directed at different things or that concepts of common defense differ from one society to another. The point is that in any society there is a concept of obligation to the society, and anyone who does not share this concept – this moral absolute – is regarded as immoral.

Finally, respect for the institutions which form the fabric of the society is a moral absolute. The institutions of society are like the steel skeleton of a building. The skeleton is an integral whole. If the institutional skeleton is weakened, the building will ultimately collapse. The same is true of society. If many of the institutions which support the society are weakened or removed by indifference or pro-active attack, the society will collapse just like the building.

Don't confuse the differences in institutions from one society to another with the need to support whatever institutions form the basis of that society. In all societies, the institutions on which they are founded must be preserved.

So we have at least four examples of moral absolutes which apply to any society wherever we find it. It is possible to argue that these examples simply show what is required for a society to exist, but that, of course, does not make them any less moral absolutes. The point is that not everything is relative, and when you parachute into any society, with respect to these four things at least, you will know right away what is right and what is wrong in that society as viewed by the people in it.

It is clear that moral absolutes exist, contrary to the view of the cultural relativist that all things are relative. The critical thing to note is that in the context of a society those moral absolutes apply to the institutions which support the society.

If support of institutions like these, which are the foundation of society, is moral, then it follows that not supporting them is immoral. This is particularly true when the institutions have reached the pinnacle of social organization in terms of the protection of the rights of individual members of society which we have achieved in this country. It is in this context that the lack of respect for and support of these institutions, which by definition is immoral, is most alarming.

There are many historical examples of societies which collapsed because their institutions were ignored or neglected. The questions raised in this chapter are whether we are heading in the same direction, and, if so, what we can do about it before it is too late.

Our Institutions

What, then, are our institutions, those things on which our society is based?

- Belief in religion in general and the morality it provides
- Belief in a federal government with limited powers which is required to operate in accordance with the Constitution
- Respect for the rule of law
- Commitment to national defense
- Belief in and promotion of high quality education by global standards
- Belief in free enterprise and free trade

Where does each of these institutions stand today in terms of its strength in support of our society? If we find that it is not strong enough, we have to determine what needs to be done to restore it to its required strength. The balance of this chapter will address the institution of religion and the state of our morality. Subsequent chapters will address each of the other institutions.

The Role of Religion

The following is an example of the attitude of President Obama and his supporters toward religion, as reported by Vice Admiral Bob Scarborough of Arlington, Virginia. Bob works with the Catch-A-Dream Foundation, which provides "hunting and fishing trips to children with life- threatening illnesses".

"As part of our annual banquet/fundraiser, we scheduled Sgt.1st class Greg Stube, a highly decorated U.S. Army Green Beret and inspirational speaker who was severely injured while deployed overseas and didn't have much of a chance for survival to come. Greg is stationed at Ft. Bragg, NC and received permission of his commanding officer to come speak at our function. Everything was on go until Obama made a policy that NO U.S. SERVICEMAN CAN SPEAK AT ANY FAITH-BASED PUBLIC EVENTS ANY MORE. Needless to say Greg had to cancel his speaking event with us."[1]

Of course, it is not clear why this particular group was considered faith-based in the first place. However, that is not the point. Inspirational speaking, which is apolitical and not faith-based per se at all, should apply to any group which thinks it could benefit from it, and faith-based groups are as legitimate as any others. The government does not have the Constitutional authority to prohibit free speech by members of the military when their speech has nothing whatever to do with faith.

Modern liberals are basically secularist at heart, but they have no business demonizing what they consider to be faith-based groups and denying them the opportunity to be as inspired and renewed by inspirational speaking as any other groups. It is particularly reprehensible that this misguided policy thinking, which is illegitimate and discriminatory, should have deprived the children of the Catch-A-Dream Foundation of inspiration which they clearly need.

This example is all too common among today's intellectual elites. They try in every way to reduce both the influence of any religion in modern society and the role which by definition religion played in our creation as a country. They are even trying to reduce reliance on the founding documents, as they define what they think is appropriate to do to modify these documents without going through the constitutionally mandated amendment process, and they are having some success.

Remember that the concept of natural rights, which was such a transformative breakthrough in the thinking about individual rights, was basically religious in nature. For the Founders, natural rights unquestionably came from a supreme being. The close connection between the two, which was so important to them, is gradually being eroded.

The Founders

Let's examine the feelings of the Founders about religion. The specific issue of separation of Church and State was discussed in Chapter 2, so we don't have to cover that here. Here are some quotations from the Founders on the subject of religion. Please note that a number of the quotations in this chapter are from a book written by Matthew Spaulding of the Heritage Foundation entitled, <u>We Still Hold These Truths.</u>

In his address to the Massachusetts militia in 1798, John Adams said, "We have no government armed in power capable of contending in human passions unbridled by morality and religion. Our constitution was made only for a moral and religious people. It is wholly inadequate for the government of any other".[2]

In an address at the Constitutional Convention in 1787, Benjamin Franklin said, "I have lived a long time, and the longer I live, the more convincing proofs I see of this truth – that God Governs in the affairs of men".[3]

"The God who gave us life, gave us liberty at the same time" Thomas Jefferson once wrote. "The hand of force may destroy, but cannot disjoin them".[4]

"Anyone of 'pious reflection', wrote Madison in Federalist 37, could not fail to perceive a finger of the Almighty hand which has been so frequently and signally extended to our relief in the critical stages of the revolution".[5]

Madison also wrote, "before any man can be considered as a member of Civil Society, he must be considered as a subject of the Governour of the Universe".[6]

According to the Founders, "all had a natural right to worship God as they chose, according to the dictates of their consciences. At the same time, the Founders upheld religion and morality – to paraphrase Washington's Farewell Address – as indispensable supports of good habits, the firmest props of the duties of citizens, and the great pillars of human happiness".[7]

In this address, Washington provided a warning which echoes down through the ages and washes over the Progressive movement, "And let us with caution indulge the supposition that morality can be maintained without religion. Whatever may be conceded to the influence of refined education on minds of peculiar structure, reason and experience both forbid us to expect that national morality can prevail in exclusion of religious principle".

According to Spaulding, "While it is often thought that religion and politics must be discussed as if they are radically different spheres, the Founders' conception of religious liberty was almost exactly the opposite. The separation of church and state *authority* actually allowed - even required – the continual influence of religion upon public life. In a nation of limited government, religion is the greatest source of the virtue and moral character required for self-rule".[8]

It is clear that in all the documents which were actually created by the Founders, in the best notes (from Madison) from the Constitutional convention, and in all their public comments, such as those in the Federalist papers, the Founders demonstrated a strong belief in a superior being. After a careful reading of these quotations, how could anyone deny that this was the case?

In fact, for them this was a necessary and sufficient condition, the sine qua non, for a view of the supremacy of natural rights compared to anything which had come before. Natural rights are rights that come with being a human being – in Jefferson's words, "Life, Liberty, and the Pursuit of Happiness".

Quite simply, (1) the Founders believed in God, (2) that belief allowed them to believe in the concept of natural rights, because without a superior being there were no natural rights, and (3) those beliefs enabled them to establish the foundation for the development of the greatest country the world has ever seen.

As Spaulding notes, the important thing to understand is that the "official separation of church doctrine and the new federal government never meant – was never intended to imply – the separation of religion and politics, or the expunging of religion from public life".[9] The Founders vision was that, if men are endowed with "natural rights", then they should be encouraged to have a relationship with the creator of those natural rights.

The Founders recognized the need for public morality and the prominent role that religion plays in nurturing morality. Spaulding points out that "The health and strength of liberty depend on the principles, standards, and morals shared by nearly all religions".[10] George Washington in his First Inaugural Address said that no nation can prosper that "disregards the external rules of order and right, which Heaven itself has ordained"

The next two quotations are both from Spaulding. "While the Constitution officially "separates" church and state at the level of doctrine and lawmaking, it also allows the general (nonsectarian) encouragement and support of religion in public laws, in official speeches and ceremonies, on public property and in public buildings, and even in public schools. Such activities were understood to be part of the free exercise of religion".[11]

"In general, the Founders saw nothing wrong with the federal government indirectly supporting religion in a nondiscriminatory and noncoercive way... Indeed, official recognition of religious faith has always been a central aspect of how we define ourselves as a political community".[12]

The problem is that "without a transcendent foundation in the 'laws of Nature and Nature's God' or by the endowment of the Creator, our rights become the arbitrary gift of government. And what government giveth, government can taketh away"[13] This is a critical point. The remarkable insight of the Founders was that each person is endowed with natural rights. These rights do not come from the government, so by definition they must come from somewhere else.

The concept of natural rights provided by a superior being is what raises the bar of individual rights to a level which has never been achieved before. It makes the rights things which are beyond the ability of human beings to change. The development of this concept was one of the crowning achievements of the 18th century.

However, if this connection between natural rights and individual rights is not vigorously defended and it is allowed to become something of little significance, then the vacuum will be filled by government definitions of rights. These rights can be changed with each change in government and exist only as long as a particular ruling government is in power. This is why totalitarian regimes, except religious theocracies, try to eliminate religion from the public square. At that point, there is no competition from religion and no concept of natural rights.

This is what Karl Marx tried to accomplish with the publication of the Communist Manifesto.

"Religion is the sigh of the oppressed creature, the heart of a heartless world, and the soul of soulless conditions. It is the opium of the people."[14] "The abolition of religion as the illusory happiness of the people is the demand for their real happiness. To call on them to give up their illusions about their condition is to call on them to give up a condition that requires illusions."[15]

So we have an unambiguous choice. On the one hand, we can believe in what the Founding Fathers believed in terms of the important role of religion and the natural rights which come from a superior being in the government and society which they were creating. On the other hand, we can believe that religion doesn't matter and that, in the view of Marx, religion is in fact a significant impediment to building the society which he was trying to build and therefore must be eliminated.

Think about what is happening today in this country to reduce and suppress the influence of religion. To the extent that happens, our society will take on more of the characteristics of Marxian society described in Chapter 4.

As suggested above, without religion and a belief in natural rights which come from a superior being, rights have to be defined by someone or something else. By definition, these would have to be the creation of man and his government and be subject to change with each change in government.

You may not believe that this could happen in America, given the commitment to natural rights which is embodied in our founding documents. However, in the next chapter, I will give you an example of a major thrust in constitutional law today to transform the Constitution from a document which provides "negative liberties" – e.g., the government cannot inflict cruel and unusual punishment on the citizens or engage in unreasonable searches and seizures, etc. – to a document which identifies "positive rights", such as the rights to affordable housing, education, food, clothing, etc., without using the specific amendment process provided in the Constitution itself.

How does this view of natural rights differ from the "Divine Right of Kings" to determine morality? Up until the founding of this country, the most current thinking was that the King had the responsibility to determine morality, because he was closest to God. The Founders took the whole idea of rights to a new level based on the concept of natural rights. It was specifically the Divine Right of Kings to determine morality against which the Founders were rebelling.

If any view other than the primacy of religion were again to prevail, it would mean that we would sink back into a much lower level of evolution of the rights of man and that the government which was created by the Founders and which has been so remarkably successful would have been at best a distraction and at worst a waste of time as viewed in the broad sweep of history.

Definition of Religion

It is important to clarify a very important point. Our government was created with a Judeo-Christian view of religion, but that is not disqualifying. The important thing was that there was a superior being from whom natural rights came ("... endowed by their Creator...").

Would the Founders have argued that this view of the superior being was the only conceivable one? I am sure that they would not, and I would like to present a more general view which I think they would embrace.

Obviously, I am not the first person to promote this view of religion. As Michael Novak has noted, "recognizing one nation 'under God' is much more important in a country as religiously diverse as America because the phrase transcends any one faith or denomination and is inclusive". Harvard professor Samuel Huntington has pointed out that "Americans tend to have a certain catholicity toward religion. All deserve respect".

For those who believe in a superior being, there are two possibilities: (1) polytheism, and (2) monotheism. Polytheism is the belief that there are multiple "gods" – this was the belief, for example, of the ancient Greeks and Romans. Monotheism is the belief in only one superior being, not many.

In today's world, polytheism would hold that the Muslim god is different from the Jewish god, which in turn is different than the Christian god, and so forth. In many cases, for example, the faith asks people to believe that: (1) they are the chosen ones, and (2) there is no hope for anyone of a different faith. This by definition leads to multiple "heavens" and "hells" existing in principle side-by-side. I find it intellectually unacceptable that there could be as many gods as there are faiths in the world.

Polytheism is very implausible, because it ends up, depending on the religion and particularly for the larger religions, condemning large portions of the human race who do not believe in that religion to second class citizens or worse. We know that this cannot be right, because we know at this stage of the evolution of human rights all human beings have natural rights, and no one has more or less of them just because of the religion they embrace.

Religion cannot at the same time be the basis for the creation of natural rights, on the one hand, and the vehicle for denying that they exist for large parts of the human race, on the other.

It is hard enough to persuade yourself that a superior being exists; it is practically impossible to believe in the multiple gods which are the essence of polytheism. What this means is that there can be only one superior being and that the gods of all the faiths in the world are one and the same, as heretical as that may seem. To put it another way, no faith has a monopoly on the correct way to have a relationship with the superior being. That would be very unlikely.

It may sound strange to believe that, for example, Allah (for the traditional Muslims), the Christian God, the Jewish God, Buddha, and "nature's god", in the words of Thomas Jefferson, are the same thing in the end, but that is the only position which makes rational sense. Perhaps that is why the different faiths generally serve the same purpose in different societies and have credos which are remarkably similar in broad strokes.

This is not to say that people should not believe strongly in a particular religion – in fact, it is necessary that they do. People find comfort in thinking about a superior being in different ways, and there is nothing illegitimate about the particular way they do it. "Freedom of religion" – the freedom to embrace any faith and to practice that faith – implies that there is more than one legitimate religion. Otherwise, the phrase has no meaning.

What I am saying is that it is very difficult to believe, to use Christian terms, that any particular religion has a monopoly on "how to get to the Kingdom of Heaven". In fact, the successful society in the minds of the Founders depended on the people following their individual faiths, whatever they are, and accepting the moral rules which are part of them. If a large segment of the population ever develops which is not a part of any religion and not an enthusiastic and active supporter of a faith, the concept of natural rights coming from a superior being is doomed.

Some of the Founders were "Deists". Franklin, in fact, described himself as a Deist in his autobiography. "Deism was not actually a formal religion, but rather was a label used loosely to describe certain religious views...Deists were characterized by a belief in God as a creator and believed only those Christian doctrines that could meet the test of reason... Nature's God was clearly the God of deism in all important ways."[16]

Thomas Paine described deism as follows: "Were man impressed as fully and as strongly as he ought to be with the belief of a God, his moral life would be regulated by the force of that belief; he would stand in awe of God and of himself, and would not do the thing that could not be concealed from either... This is deism."[17]

It is interesting to note that Franklin on June 28, 1787 went so far as to make a formal motion for prayers at the Constitutional Convention, as follows:

'I therefore beg leave to move, That henceforth Prayers, Imploring the Assistance of Heaven and its Blessing on our Deliberations, be held in the Assembly every morning before we proceed to Business, and that one or more of the Clergy of this city be requested to officiate in that Service.... The Convention, except for three or four persons, thought prayers unnecessary."[18]

Deism attracted interest as late as the time of John Quincy Adams. However, most Americans continued to believe in one of the traditional, organized religions, and deism quickly died out.

Deists clearly believed in a creator and some form of religion related to a creator, as did the more traditional Founders. In fact, all of the Founders believed in a creator – there was not an atheist among them – and this commonality of views underscores their fundamental belief in something new. This something was natural rights, and their beliefs led to a form of government designed to protect those rights.

The bottom line is that a belief in a superior being leads directly to the concept of natural rights and to the highest level of individual rights which the evolution of human rights has yet achieved. It is plausible to argue that natural rights could simply spring from the ground or the minds of men. However, in that case, beauty is in the eye of the beholder, and, since such an assessment is very subjective, what is beautiful will change over time. If this happens, the cultural relativists will have triumphed.

Objections to Religion

What about evolution? Doesn't the belief in evolution, which I share, make a supreme being impossible to conceive? I don't think it does. What it does do is make any concept other than a monotheistic superior being inconceivable. In principle and intellectually, there is not necessarily a conflict between evolution and the belief in a universalist supreme being.

There is no question that a belief in evolution makes it difficult to support the belief of some religions that the earth was created all at once a relatively short time ago. The practical evidence that the earth as we see it today has gone through convulsions which have taken millions of years to take place is only consistent with creationism if somehow all of what we find on earth today was created to look like it had been here for a long time. That is a proposition which it is hard to support.

However, the idea that evolution and a religion which is not time-limited are incompatible still cannot be dismissed. It is apparently true that the universe as we know it was created by a "Big Bang". The problem is what is the answer to the question, "what created the Big Bang?".

It is admittedly an intellectual stretch to consider that the Big Bang was instituted by a superior being and that this superior being played some role in evolution itself, but it is not out of the question. If religion or belief in a superior being cannot be ruled out by any scientific techniques we have available to us today, then certain other considerations need to be examined.

After all, religion is a supreme act of faith. There is very little scientific evidence which can be mustered to prove that a superior being exists. However, the important point is that the scientific evidence is totally incapable of proving that a superior being does not exist. Therefore, it follows that the two ideas of evolution and a superior being must be viewed as not incompatible, particularly if it is possible that they come together at the time of the Big Bang.

We can see now that a belief in a superior being which is the basis for the natural rights on which our country was founded cannot be rejected out of hand. Until and if it can be, religion as the source of natural rights which so transformed and elevated the concept of individual rights in this country must be supported. Since religion is one of the institutions of our society outlined above, and support of those institutions is moral, it follows that not supporting religion as a source of natural rights is immoral.

This is the problem for agnostics and atheists. Freedom of religion means that it is ok to hold their views. What is not ok is to try to tear down a whole societal edifice which has been created on the basis of a monotheistic supreme being from whom come natural rights and which gives one the freedom to be an atheist or an agnostic in the first place.

National defense is one of the first responsibilities of any government, and its purpose is to keep safe all of its citizens. For an atheist or agnostic to tear down the natural rights edifice because of a belief that a supreme being does not exist would be like trying to sabotage our national defense effort because the saboteurs don't believe in war. Doing so is illogical and self-destructive.

In fact, as pointed out by none other than Charles Darwin himself in the last sentence of his great book, <u>On the Origin of Species,</u>

"There is grandeur in this view of life; with its several powers having been originally breathed by the Creator into a few forms or into one; and that, whilst this planet has gone cycling on according to the fixed law of gravity, from so simple a beginning endless forms most wonderful and most beautiful have been, and are being evolved."

There are, of course, many experts on both sides of the issue of evolution versus a supreme being, and the general impression is that scientists can only believe in evolution by definition. One of my favorite books on this subject is by Kenneth R. Miller, who is a cell biologist and professor of biology at Brown University. In his very thought-provoking book, Finding Darwin's God, his conclusion is the following: "What kind of God do I believe in? The answer is in [Darwin's own words]. I believe in Darwin's God" (see Darwin's quote above). So do I.

There are some people who raise another concern about religion. They assert that "there have been more deaths over recorded history from religious wars than from non-religious wars", and they use this claim to dismiss religion as illegitimate. Anything which would have this kind of negative impact simply could not be an essential or even a positive force.

In the first place, the data to support this assertion are sketchy. In addition, there are definitional considerations – are we talking about all of recorded history, or just the 20th century? Finally, wars are fought for a number of reasons – power, territory, defense, and, yes, religion. It is very hard to sort out which wars were fought for what reasons, given the long historical time periods and the paucity of the data.

Nevertheless, let's grant the assertion. Does it really matter in the first place? Individual societies feel strongly about their religions, given their central place in the societies, and it is not surprising that societies fight with each other because of religion regularly and at great human cost.

However, this has nothing to do with the role of religion in each society, and this is what we are talking about with the development of natural rights. It could be, and probably is, the case that religion plays exactly the same role in each of the societies at war because of religion. The war is not fought about the legitimacy of religion in society but about the differences between particular religions.

In the end, wars resulting from differences in religion are tragic if you believe in the monotheistic God described above. The religious differences really don't matter.

God in America

Newt Gingrich has written an excellent little book, entitled <u>Rediscovering God in America</u>, which takes readers on a walking tour in Washington, D. C. of all the significant monuments and memorials, including the Supreme Court, the Jefferson Memorial, the Lincoln Memorial, the Capitol, the Roosevelt Memorial, the Kennedy Memorial, the Library of Congress, and the National Archives. What he finds is constant references to God, as the following examples and quotations illustrate:

- Supreme Court

   All sessions begin with the Marshall of the Court announcing, "God save the United States and this honorable court". "This phrase has been used for almost two hundred years. It was not adopted as a ceremonial phrase of no meaning; it was adopted because justices in the 1820s wanted to call on God to save the United States and the Court."

   Justice William O. Douglas wrote in an opinion in 1952, "We are a religious people, and our institutions presuppose a supreme being."

- Jefferson Memorial

   Around the chamber on the interior dome is the quotation, "I have sworn upon the altar of God, eternal hostility against every form of tyranny over the minds of men"

   On Panel One inside the dome are the immortal words from the Declaration of Independence, "We hold these truths to be self-evident: That all men are created equal, they are endowed by their Creator with certain unalienable rights, that among these are life, liberty, and the pursuit of happiness".

- Lincoln Memorial

The Gettysburg address is inscribed in the monument. A portion of it reads, 'We here highly resolve that these dead shall not have died in vain, that this nation, under God, shall have a new birth of freedom; and that government of the people, by the people, for the people, shall not perish from the earth."

- The Capitol

The House and Senate open their daily sessions with the Pledge of Allegiance.

- Roosevelt Memorial

On D-Day, Franklin Roosevelt led the nation in prayer in a nationwide radio address, saying the following, "Almighty God, our sons, pride of our nation, this day have set upon a mighty endeavor, a struggle to preserve our republic, our religion, and our civilizations, and to set free a suffering humanity."

- Kennedy Memorial

Inscribed on the memorial is John Kennedy's inaugural address in 1961 the conclusion of which is "With a good conscience our only sure reward, with history the final judge of our deeds, let us go forth to lead the land we love, asking His blessing and His help but knowing that here on earth, God's work must truly be our own."

No one can deny that these immortal words which were the foundation of our form of government have played a critical part in the development of our country over the last 220 years. Not only did the Founders underscore the importance of God, but some of the most recent presidents have embraced the same theme. It is no wonder that these words have been literally carved into stone to remind future generations about where we came from and who we are.

Polling data show how much the American people fundamentally stand with the Founders. NBC recently asked people the following question, "Do you believe that the word 'God' should stay in American culture?" They had the highest number of responses they have ever had to one of their polls. 86% supported keeping "In God We Trust" and "God" in the Pledge of Allegiance.[20] Another poll by CNSnews.com shows that 92% of Americans believe in God.[21] Only 5% oppose a National Day of Prayer, even though two weeks earlier a federal judge in Wisconsin ruled that the Day was unconstitutional.

Is it conceivable that an atheist from California, who wants to remove the words "under God" from the Pledge of Allegiance and "In God We Trust" from our coins, would be allowed to remove all mentions of God in these inscriptions or activities and thus erase the philosophical foundation of our country? This is the epitome of the Tyranny of the Minority which I discussed earlier, and it can't be allowed to happen.

Americans should celebrate these magnificent monuments, the words which are inscribed on them, and the words which are used to open the proceedings of two of the three branches of government created by the Founding Fathers and promote both to their children and to immigrants who come to this country.

Current Trends

It is clear that the syllogism outlined above - God, natural rights, and country – is the heart of the philosophy of the Founders. "The Founders held to this simple syllogism: morality is necessary for republican government, religion is necessary for morality; therefore, religion is necessary for republican government"[21] Not only was this their firm conviction, it was also the way they expected the American people to think about what they had created and why.

Unfortunately, it would follow that as and if the leaders of the country and the people as a whole began to move away from this philosophy by becoming indifferent to organized religion of any kind and its central role in the ongoing success of this country, some things would start to change. Rights would be thought of more as things bestowed by government, as opposed to being bestowed by a supreme being.

It is therefore possible that in this environment people can believe in anything or nothing at all, and there is no philosophical foundation for whatever rights people are deemed to have. This should show up as a moral malaise in which people are inclined to: think they are better than they are, be selfish and inner-focused, and think more about doing things which even the current society feels is wrong if they think they can get away with it.

The evidence of a moral malaise would include such social trends as: (1) a decline in church-going, (2) a rise in births out of wedlock, and (3) an increase in abortions. It is unlikely that the moral malaise in the US is an isolated phenomenon. It must be the case that these trends are in evidence in developed countries in Europe, which have had a similar religious history to ours. What is the evidence?

Church Attendance

Happily, "contrary to common belief and despite a sexual abuse scandal that has engulfed the Catholic Church, attendance at churches in the United States has declined only slightly in recent decades"[22], based on a study of data through 2006 reported by Reuters on 4/15/10. "We are an exception. We attend religious services and hold religious beliefs far more than most people in other well-to-do nations"[23], according to Philip Schwadel, a sociologist at the University of Nebraska-Lincoln.

However, at the margin, things are changing. Based on a survey by Lifeway Research, a Christian research firm, covering 1,200 18-29 year-olds of all faiths "Most young adults don't pray and don't worship... 65% rarely or never attend worship services. In the group's survey, 72% say that they're 'really more spiritual than religious"[24].

The data for young adults are very disturbing. For them, the view of the Founding Fathers that a superior being creates natural rights and that a government created to protect those rights should have specifically limited powers is increasingly irrelevant. As a result, anything goes and is worth trying.

What we see in this country is mirrored in Western Europe. "I don't go to church, and I don't know one person who does", says Brian Kenney, 39, who is studying psychotherapy and counseling at Dublin Business School. "Fifteen years ago, I didn't know one person who didn't."[25] According to the Dublin Archdiocese, church attendance in Ireland has fell from about 85% in 1975 to 60% in 2004, and it seems that there is no end in sight.[26]

"Ireland is not an exception. Every major religion except Islam is declining in Western Europe...the drop is most evident in France, Sweden, and the Netherlands, where church attendance is less than 10% in some areas."[27] In fact, based on the latest available data, the percentage of the population which attends church regularly is 21% in France, 4% in Sweden, and 35% in the Netherlands[28]. The US percentage is 44, the highest for any developed country. In the UK, church attendance has fallen from an already abysmal 11% in 1980 to about 6% in 2010.

The dramatic decline in church-going is a symptom of the malaise, and it has real world implications. For example, "The biggest single consequence of the declining role of the church is the huge decline in fertility rates", says Ronald Inglehart, director of the World Values survey, a Swedish-based group that tracks church attendance. [As a result], Germany and France, [for example], won't be able to maintain the generous welfare programs that have given most workers a lifetime of economic security."[29]

Social mores are changing significantly and in a direction which promotes self-determination and an environment in which anything goes. As I indicate above, if no one agrees on what is wrong, then by definition everything is right. This represents a 180 degree turn from the syllogism of the American Founders, and the results are predictable.

"Many Europeans may possess individual spirituality, but it may no longer be expressed in traditional religion."[30] "Bob Kenney, the Dublin student [quoted above], says he's merely typical of his generation. "I'm very spiritual", he says. "I speak to an energy force I call God, and I get answers," he says. "If you can get a spiritual connection without going to church, why go to church?[31]

This atomization of the concept of religion, where each person defines religion in his/her own way, makes it impossible for people to maintain a commitment to and respect for the syllogism of the American Founders. It is easy to see why there is increasingly no moral compass, because morality comes from a generally accepted belief in a superior being.

Births Out of Wedlock

The percentage of births to unmarried women in the US has risen to 41% in 2008 from only 18% in 1980. Four out of every ten babies do not have a traditional family when they are born. In fact, "the percentage of births to unmarried mothers is increasing worldwide, according to a new federal report that shows a universal upward trend over the last 25 years. 60% of births to women ages 20-24 were non-marital in 2007, up from 52% in 2002."[32]

"Countries with a higher proportion of births to unmarried mothers include Iceland, Sweden, Norway, France, Denmark and the United Kingdom: countries with a lower percentage than the USA include Ireland, Germany, Canada, Spain, Italy, and Japan. The Netherlands has increased ten-fold from 4% in 1980."[33]

Clearly, the "new morality" is what you would expect to see if a moral malaise were taking place. While I am obviously not a sociologist, I think most would say, based on evidence since the beginning of time, that a family unit is the most productive for child-raising. In terms of the large volume of unmarried births and the high percentage of live births which unmarried births represent, we see people doing whatever works for them, as opposed to what would be best for their children.

Why does it matter? According to the Heritage Foundation, "research shows that a child raised in a home where Dad is married to Mom is much less likely to live in poverty, get arrested as a juvenile, be suspended or expelled from school, be treated for emotional or behavioral problems, or drop out before completing high school".[34]

In view of this kind of research, it seems logical for government to promote marriage. In fact, President Obama agrees. "...five years ago, he wrote: 'in light of these facts, policies that strengthen marriage for those who choose it and that discourage unintended births outside of marriage are sensible goals to pursue."[35]

What actually happened? "Rather than adopt policies to reverse the 50-year spike in births outside marriage, President Obama in his 2011 budget would eliminate the one program dedicated to encouraging healthy marriage. In its place would be a program promoting a notion of 'fatherhood', that doesn't involve the father being married or in the home."[36]

This inconsistency and total disregard for the best interests of the children born out-of-wedlock, is outrageous. This situation is sapping the long-term vitality of the American Dream for millions of our children and for future generations.

Abortion

The bottom line for the US, based on data through 2005, is that since the Supreme Court decision in 1973 in Roe v Wade there have been over 45 million legal abortions performed, an average of about 1.1 million per year.[37] This is really an extraordinary number. Obviously, there is a variety of reasons why women get abortions, and not all of them are arbitrary.

However, it is important to consider the impact on the country as a whole of such a large amount of abortions. The population of the US is currently about 310 million. At the extreme, if all of these abortions had not taken place, the US population would be over 350 million, and perhaps half, or over 20 million, of those whose lives were aborted would be in the workforce. The workforce is currently about 150 million, and the addition of these 20 million to the workforce would increase it by about 13%.

Long-term economic growth is a function of two things: (1) increases in the workforce and (2) the growth rate of productivity of the workforce. Productivity growth is fairly constant over time, but the workforce growth rate under these circumstances would have been significantly higher. Therefore, economic growth would have been greater than it has been.

I understand that abortion is an intensely personal issue, and no one should make judgments about a woman's decision about it. However, the combination of these intensely personal decisions in total does have a social cost.

The most important consideration is not economic but moral. It is likely true that most major religions do not tolerate abortion. As a result, through the 1960s, abortion was discouraged by the church and by families, even if there was some hardship for the pregnant woman. As far back as 1970, abortions in the US were about 190,000.[38]

However, the number of abortions increased in the late 1970s and 1980s to a peak level of almost 1.5 million in 1990, an increase of 8 times. Even considering the increase in women of child-bearing age, this is a dramatic increase. Since then, the number of abortions has declined to a constant level of about 800,000 in the early years of the 2000s. This level is still 4 times the level in 1971.

While the number of abortions has declined, the fact that there are so many abortions at all, compared to what was the case 35 years ago, is a demonstration of the decline in morality and the permissiveness of current social standards in this country.

What is the situation in other countries? "The abortion rate varies widely across the countries in which legal abortion is generally available and has declined in many countries since the mid-1990s."[39] The abortion rate in the US – the number of abortions per 1,000 women ages 15-44 per year – at over 20 is higher than for any developed country in the world.[40] The Netherlands has the lowest rate at about 5, while Vietnam has the highest rate at over 80.

Conclusion

The bottom line is that personal and cultural morality in our society stem from the concept of a superior being and the natural rights, form of government, and institutions which are part of it. It follows that the different religions which provide this kind of moral structure in the development of the rights of man should be respected and promoted. Not doing so is immoral.

As one of the key institutions which were outlined above as critical for the maintenance of our society, religion is in retreat. With it, in turn, the concept of natural rights which has elevated our society to a new pinnacle in the organization of any society, is losing its force. As the legs of the structure which holds up our society, as described above, are removed one-by-one, America as the great last hope for the world will be in an irreversible decline.

What can be done? Recall that I said at the beginning of this Chapter that criticism and commentary are not sufficient to persuade a rational person. One has to suggest a plan of action to address the issue or person being criticized.

Here is what I propose:

- Stress in all public communications by public officials and ask ordinary citizens to stress in their private communications the critical relationship between belief in a superior being, the natural rights which come from the superior being, and the government which was built on these religious convictions.
- Promote the idea of a universal religion to avoid irrelevant ideological arguments.
- Put religion back onto Main Street where it belongs.
- Do not tolerate tyranny of the minority.

Remember the words of John Winthrop, Governor of the Massachusetts Bay Colony, who wrote on the Arabella during the voyage to Massachusetts:

"We must consider that we shall be as a City upon a hill. The eyes of all people are upon us. Soe that if we shall deal falsely with our God in this work we have undertaken, and so cause him to withdraw his present help from us, we shall be made a story and a byword throughout the world".[41]

Bottom Line

- Moral absolutism based on religion is the foundation of our country.
- Cultural relativism is by definition secular in nature and not a permanent foundation for social morality. Under cultural relativism, morality can be whatever you want it to be.
- The syllogism of the Founders was a superior being, natural rights, and constitutional government.
- A universalist view of religion bridges the gap between the Judeo-Christian background of the Founders and all the present day religions.
- Trends in church attendance, births to unwed mothers, and the number and rate of abortions are an indication of a moral malaise which threatens one of our key institutions.
- Specific actions can be identified to reverse the decline in the role of religion and avoid falling back into a society in which what is right or wrong is either indeterminate or arbitrary.

CHAPTER 4
LIMITED GOVERNMENT

A government big enough to give you everything you want is a government big enough to take from you everything you have.

President Gerald Ford

The first of the institutions on which the structure of our society depends – Religion and Morality – was discussed in Chapter 3. The next one on that list – Limited Government - will be discussed in this Chapter.

It is quite apparent to everyone that respect for government at the local, state, and national levels is very low. For example,

- Those who have a favorable rating of the US Congress represent only about 20% of the total population.
- With some exceptions, state governments have not been able to control spending, allowing it to increase at a rate far in excess of population growth. The result is chronic deficits and higher taxes. Spending is difficult to control because of the demand for higher wages by public employee unions, the almost unbelievable costs of benefits promised to these unions, and an entitlement mentality which produces a plethora of ill-conceived and never-ending programs designed to promote equality of condition as opposed to equality of opportunity

- Voters are generally apathetic. Turnout is very low in primaries and not much better in actual state and local elections. In presidential elections in particular, usually less than half of those eligible to vote actually cast a ballot. A large number of voters don't care, because they don't see a direct link between their vote and actual results. Remember Jefferson's quote from Chapter I - "If a nation expects to be ignorant and free, in a state of civilization, it expects what never was and never will be."

   If voters are not informed and not involved, they will get the government which they deserve, and their representatives will tend to act more and more on their own rather than being aware of and responsive to the views of those who elected them. Jefferson's point is that in this environment, they will gradually lose their freedoms.

- The Congress is appallingly dysfunctional. There is none of the bipartisanship, which all important legislation requires, according to Daniel Patrick Moynihan, former Senator from New York, and the result is gridlock. In fact, many voters are hoping for exactly that when they vote. They would rather have the government make no decisions at all if it is not going to make them based on a thorough vetting of all points of view about what is in the best interest of the country.
- There has been so much scandal and corruption on both sides of the aisle in Congress that voters are disgusted with politicians, not only collectively, but as individuals. The oversight bodies in both Houses of Congress don't seem able to set and enforce standards which ordinary voters can understand and support. So the nightmare of politicians defying any rules and doing what they want to do goes on.
- Both national parties when they have been in power have operated with enormous deficits which represent a heavy burden on future generations, because they could not control spending. A significant amount of the excess spending is related to the entitlement programs discussed in Chapter 2, and there is every indication that the cost of these entitlement programs is going to increase enormously in future years, unless dramatic action is taken. The situation is similar in many states.

There are two possible reasons why so many people are disenchanted with government: (1) what the government is doing is what it is supposed to do, but it is not being done well, or (2) government is doing far more than it should if it is truly limited by the Constitution, or both. It is not surprising that the government is doing most of what it is currently doing badly. As a result, for this reason, at least, the people of this country are justifiably concerned.

However, what is of critical concern at this time in our history is that the government is in fact doing far more than it is authorized to do under the Constitution. If this is the case, it follows that the government has extended its influence into our daily lives, by definition, illegitimately. The government's powers were limited by the Founders, because they did not believe that any government should have broader powers than they were willing to grant.

There were presumably two reasons for this: (1) government powers should be limited to maximize liberty (see Chapter 5 – the more government involvement in the society the less liberty there is and vice versa), and (2) it is unlikely that the government by definition would be able to manage successfully more than a very limited number of activities, if any. It was much better to rely on the states for those powers not enumerated. Both of these concerns of the Founders are borne out in the above list.

The constant accumulation of powers by government is very subtle, and it is not clear as people go through life year-to-year exactly what is happening. It is important to recognize that this accumulation is insidious, to keep vigilant every single day for new government activities and the resulting reductions in liberty, and to both stem the tide and begin to roll back the illegitimate governmental activities.

How far have we departed from the concept of limited government which promotes individual liberty and free choice and which is one of the foundations of the American social contract? To determine this, let's consider the polar opposite of the government created by the Founders. That would be a command and control system of government in which all activity is controlled by a small group of elites. The best example of this is the Soviet Union and its inevitable intellectual bankruptcy.

Consider what Karl Marx had to say about Communism as a form of government. According to the <u>Communist Manifesto</u>, in paraphrase, Communism has ten essential planks:

- Abolition of private property
- Heavy progressive income tax
- Abolition of rights of inheritance
- Confiscation of property of emigrants and rebels
- State control of credit
- Government ownership of communication and transportation
- Government ownership of factories and instruments of production
- Equal liability of all to labor
- Corporate farms and regional planning
- Government control of education

What was the intended result of achieving these ten planks? Marx was trying to create an environment of equality of condition by redistributing resources and capital, as opposed to equality of opportunity. These two are diametrically opposed views of how a society should be organized.

If a government has equality of condition as its primary objective, there is, by definition, no equality of opportunity. Citizens are not allowed to make of their lives what they choose, because, among other things, if they are too successful, the fruits of their success will be appropriated in one way or another and transferred to citizens who have not been as successful. Constant redistribution of wealth is a necessary condition for equality of condition to be realized.

If, on the other hand, a government has equality of opportunity as its primary objective, the result will be an environment in which each person can achieve his or her dreams to their full potential. Because people have vastly different capabilities and experience good fortune to varying degrees, equality of opportunity will, by definition, produce inequality of outcomes.

It is certainly true that in this environment some people will fall through the cracks, and so a safety net of a limited extent is a necessary part of a society focused on equality of opportunity. It is interesting to note, however, that the people who do fall on hard times do not stay very long in that condition in an environment in which there is equality of opportunity, unless government programs are implemented to "help" them and, in a triumph of unintended consequences, fail to do so by keeping them dependent.

Over the years, societies which have promoted equality of condition have been miserable failures from the standpoints of economic growth, social policy, national defense, recognizing and preserving the rights of individual citizens, etc. "The practical results of Communism have been horror and atrocity for those under Communist rule"[1]. There is a large number of societies today which have given up pursuing equality of condition.

This is because they have realized from their own experience, in some cases, and the experiences of others, in other cases, that equality of opportunity and the inequality of outcomes it produces represent the best way to create a high level of national prosperity and individual liberty. All of them, to one degree or another, have introduced elements of equality of opportunity, with predictable results. The best example today might be China.

If we in this country adopt any of the planks of the Communist Manifesto, it is clear that we will be moving in the direction of a type of government for which there is not one single example of success, however you define it. The Western European countries have been on this path for some time, but even they are beginning to realize, from their own experiences, that equality of condition is a failed concept.

There comes a time when societies focused on equality of condition cannot afford what they have promised, even if it made sense to them at some point, and to survive as countries they have to introduce elements of equality of opportunity. A society cannot stand when the increasingly small productive part of the economy is asked to pay for the growing entitlements of the unproductive part. In a society focused on equality of opportunity, all the people are productive to varying degrees.

It is instructive to review these planks from time-to-time, because they provide a yardstick with which to measure the movement of our own government over time. Where do we stand in terms of Marx's 10 planks?

- <u>Abolition of Private Property</u> – Assuming control of major sectors of the economy without just compensation for the investors (the automobile industry, for example) is the clearest demonstration of the direction in which the Obama administration is going.

    It is important to note that when General Motors was reorganized by the Obama Administration two things happened: (1) the bond holders, who had invested in the company in part because of legal protections which made them, by definition, senior creditors, were immediately turned into junior creditors with a smaller part of the company as security, and (2) the labor unions who had no significant investment in the capital structure emerged in a position which is senior to that the original bondholders.

    This was unquestionably an expropriation of assets <u>a la</u> Karl Marx, but it is not yet typical.

- <u>Heavy Progressive Income Tax</u> – Not only is the income tax schedule in this country highly progressive, but half the people don't pay any taxes at all. How long can you tax the productive sector of the economy to pay for equality of condition?
- <u>Abolition of Rights of Inheritance</u> – The estate tax in this country is already at confiscatory levels, and this Administration is determined to keep it that way. The rationale seems to be a redistributive one, once

again, because successful people who accumulate some wealth have a significant portion of that wealth taxed away, which prevents them from passing on to future generations a significant portion of the fruits of their labors.

- <u>Confiscation of Property of Emigrants and Rebels</u> – Based on legislation passed by the Clinton administration, emigrants from the US are taxed on their assets before they leave the country as if they had died. Therefore, many people have to pay significant taxes before they leave.
- <u>State Control of Credit</u> – We have an independent Federal Reserve, and therefore this plank does not apply.
- <u>Government Ownership Of Communication and Transportation</u> – Efforts by the Obama administration to control large portions of our electronic communications (mandating equal time for liberal shows on radio stations and restricting the time for conservative talk shows, as an example, even though the evidence is clear that people don't listen to the liberal shows and support in great numbers the conservative ones) are increasing, and it is not clear what the outcome will be. What the administration is pursuing is what is called the "Fairness Doctrine".
- <u>Government Ownership of Factories and Instruments of Production</u> – See the comments above about the auto industry.
- <u>Equal Liability of All to Labor</u> – Marx believed that the working class (the "proletariat" (the working class of capitalism) had been exploited by the "bourgeoisie" (upper/middle class), and it was thus important to do whatever was necessary to support the working class. Specifically, he wanted to eliminate economic classes by redistributing resources and capital.

The "industrial armies' to which Marx referred are today's unions. Much has been done to support labor in general and unions in particular since the late 19th century, but some of the abuses of this kind of support were described in Chapter 2. The teachers unions may, in fact, be a ruling proletariat, but we are far from the world which Marx envisioned.

- <u>Corporate Farms and Regional Planning</u> – This does not seem to be a factor in today's economy.

- <u>Government Control of Education</u> – Education in this country has historically been provided by the states, based on standards they themselves create. As will be discussed in Chapter 8, educational achievement in this country is very poor relative to that of other developed nations, so that creates an opportunity for the federal government to step in to try to "fix" the problems with mandates, incentives, and penalties for state education and thus to control education.

To be sure, we are not a communist country, and we never will be. The point is that any movement toward the achievement of any of these planks is dangerous in a long-term sense and must be reversed. It is not clear why any movements away from the institutions which are supposed to support this society are tolerated in this country in the first place. They must be identified for what they are, by reference, for example, to the blueprint of Marx for a communist society.

Many people may not realize the basic direction in which the Obama administration is moving. It is proceeding with dazzling rapidity in the direction of a command and control economy and away from the founding principles. The Tea Party movement is an encouraging demonstration of the response of the American people when they do see what is happening.

The bottom line is that voters are disgusted and fed up with the lack of political professionalism in Washington, the lack of support for the institutions which underlie our representative democracy, and the resulting drift toward equality of condition. The only recourse to preserve our representative democracy is to engage in a Second Revolution to overthrow the establishment in Washington, replace it with elected officials who meet the standards outlined at the end of Chapter 1, and Restore the Future for our children and grandchildren.

The Revolution for which I am calling has nothing to do with ideology or political party. Both parties are guilty. Rather, it has everything to do with restoring representative government to the Constitutional standards envisioned by the Founders, so that the business of the country can be effectively conducted, equality of opportunity maintained, and the nation preserved.

The present lack of faith in one of the key structural elements of the organization of our society – Constitutional government - is a very bad omen for preserving it.

Remember that one of the hallmarks of individual liberty which the Founders worked so hard to protect is limited government. The more government that there is the less individual liberty there is and vice versa. Think back to Thoreau's comment at the beginning of Chapter 1.

Let's look at the facts about government spending today. Where does our money go? Figure 4-1[2] shows how much money as a percentage of the budget in fiscal year 2012 was supposed to be spent on each activity.

Figure 4-1

More Than Half of the President's Budget Would Be Spent on Entitlement Programs

In combination with other entitlements, Medicare, Medicaid, and Social Security constitute the lion's share of President Obama's 2012 budget. In contrast, spending on foreign aid represents 2 percent.

PERCENTAGE OF THE PRESIDENT'S FY2012 BUDGET

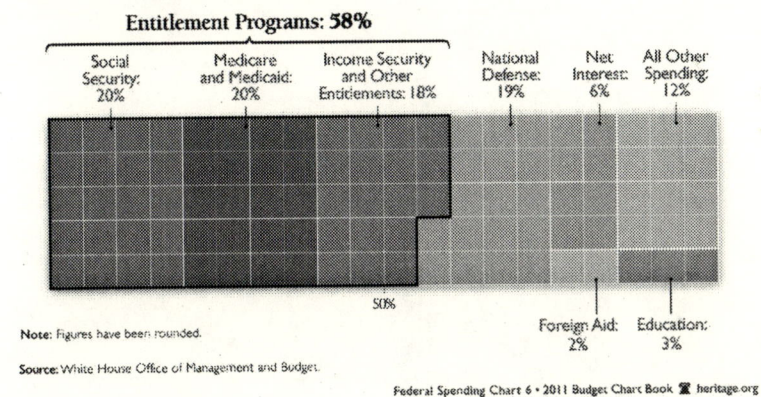

Entitlement Programs: 58%

- Social Security: 20%
- Medicare and Medicaid: 20%
- Income Security and Other Entitlements: 18%
- National Defense: 19%
- Net Interest: 6%
- All Other Spending: 12%
- Foreign Aid: 2%
- Education: 3%

Note: Figures have been rounded.
Source: White House Office of Management and Budget.

Federal Spending Chart 6 • 2011 Budget Chart Book • heritage.org

It takes only a quick glance at the chart to see how monstrously distorted our spending is. Consider the following:

- One of the primary responsibilities of the Federal government is national defense. The amount to be spent on defense is only 19% of the budget. Chapter 7 addresses the whole issue of national defense and the status of our readiness.

- Entitlement spending (that is, spending on Social Security, Medicare, and Medicaid) is supposed to be 58% of the budget. This is an absurdly large number, and it has the effect of crowding everything else out, including defense. Because these programs are inviolable and on auto pilot, it would be difficult to spend more on national defense even if that were necessary.

Figure 4-1 shows that the government's expenditures for Social Security are part of the budget. Remember that Social Security is designed to be self-funding, so that from a funding standpoint, as opposed to a spending standpoint, it is not as much of a burden as indicated in Figure 4-1. However, in terms of spending, it is still an important part of the total governmental influence on the economy.

Figure 4-2

Mandatory Spending Has Increased Five Times Faster Than Discretionary Spending

Only one-third of the federal budget, discretionary spending, is subject to annual budgets. The remainder, mandatory spending, is set on autopilot without congressional debate and has increased more than five times faster than discretionary spending. Most of the current increase is due to entitlement spending.

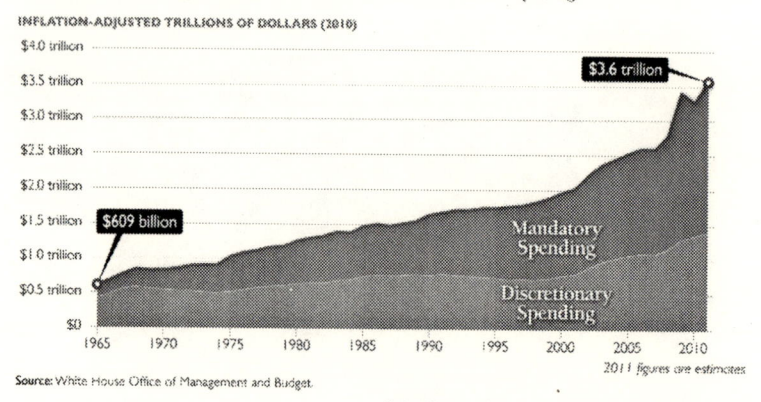

INFLATION-ADJUSTED TRILLIONS OF DOLLARS (2010)

[Chart showing Mandatory Spending rising from $609 billion in 1965 to $3.6 trillion in 2010, with Discretionary Spending shown below. 2011 figures are estimates.]

Source: White House Office of Management and Budget.

Federal Spending Chart 4 • 2011 Budget Chart Book • heritage.org

- Entitlement spending is not the only problem. As shown in Figure 4-2[3], the budget can be divided into two broad categories: (1) mandatory spending, which is spending on entitlements and which is not subject to annual budgets, and (2) discretionary spending, which is everything else and what budget battles are typically about.

It is irresponsible to focus attention on just 40% of the budget and totally ignore the 60%, which is spending on entitlements. You can see that entitlement spending has grown at a rate over the last 45 years which is 5 times as fast as the rate of growth of discretionary spending, and it appears that the rate of growth is accelerating.

There are at least two major problems with this situation: (1) entitlement spending is not provided for in the Constitution, except under the most liberal interpretation, and is therefore unconstitutional, and (2) it demonstrates more than anything else how upside down the Government's role in our economic life really is – the government should be trying to promote economic activity, so that the exercise of personal liberty will create a situation in which a rising tide lifts all boats.

Instead of doing this, the Government is doing just the opposite. It is implementing an equality of condition strategy by taxing the productive sector of the economy to pay benefits to the unproductive sector. In the extreme, this strategy would collapse of its own weight, because the productive sector would gradually be eliminated, and there would be nothing left to tax and no revenues to support the legislatively authorized entitlement programs.

- This is exactly the situation described in Ayn Rand's <u>Atlas Shrugged.</u>

Burdened with a mountain of regulations, massive government interference in resource allocation, government price determination in the economy, and punitive taxes, the major industrialists, who were responsible for what economic growth there could possibly be, went on strike and walked away from their businesses.

As a result, the economy collapsed, the country entered a depression, and the government bureaucrats, those who were responsible for creating the situation in the first place, were destroyed. They were helpless to do anything about it, because they did not have a private sector on which to rely.

These entrepreneurs all concluded correctly that the best thing to do for the country was to drop out and accelerate the process of decline. They gathered one-by-one in a place which could not be located and bided their time until the collapse of the economy and the resulting depression were so devastating that they would have the unrestricted opportunity to come back and reinstitute their particular contributions to economic growth and prosperity.

With almost 60% of Federal Government spending today on entitlement programs, with no end in sight, we may shortly reach a tipping point at which time there is no turning back, and the hypothetical world of <u>Atlas Shrugged</u> becomes a reality.

Economics

Fiscal Soundness

It is necessary to examine the present state of our financial affairs to see if our current situation impinges significantly on personal liberty. Consider the following:

- The Democratically controlled Congress in the fall of 2010 for the first time in many years did not produce a budget for the current year – FY 2011. It was only in the spring of 2011, after the Republicans took over the House in the November 2010 elections, that a budget for FY 2011 was actually created, and that was about half way into the fiscal year itself.

How can the Congress be allowed by the electorate to behave in a contrary way to what they have to do for themselves? The reason has to do with the politics in an election year of not wanting to discuss enormous deficits and huge additions to the national debt. The only explanation for this behavior has to be that the Democrats were arrogant enough to think they could get away with it and do something which was unprecedented.

- It is little comfort that when a budget for FY 2011 was finally created, it called for a deficit of $1.4 trillion.

Figure 4-3

National Debt Set to Skyrocket

In the past, wars and the Great Depression contributed to rapid but temporary increases in the national debt. Over the next few decades, runaway spending on Medicare, Medicaid, and Social Security will drive the debt to unsustainable levels.

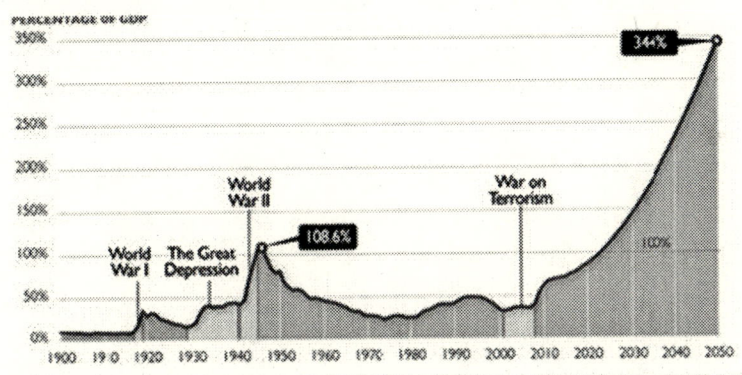

Source: Heritage Foundation calculations based on data from the U.S. Department of the Treasury, Institute for the Measurement of Worth, Congressional Budget Office, and White House Office of Management and Budget.

Debt and Deficits Chart 1 • 2011 Budget Chart Book ⚡ heritage.org

What are the implications for the national debt of the United States?

Figure 4-3[4] shows the national debt as a percent of Gross Domestic Product (GDP) – which is the value of all the goods and services in the country produced in a year - at the end of each year since 1900, with projections out to 2050. See the explanation below of the importance of this relationship.

It can be easily seen that the national debt as a percentage of GDP has increased dramatically in the last few years under the Obama administration. What is far worse is that in the budgets which Obama submitted to the Congress for FY 2012 and beyond, trillion dollar deficits and additions to the national debt were proposed for the next ten years.

What is perhaps the most striking feature of Figure 4-3 is that if present trends continue the total national debt will be 344% of GDP as soon as

2050. This is such an astronomic level that it would never be achieved in fact, because the country would default. No one would lend to the government if this situation existed.

Figures like these should be circulated to and well-known by the American people. I don't believe that they know how bad things really are. The figures are unambiguous, and they should serve as the basis for active dialogue and participation by all Americans, because this is what will actually turn the tide and Restore the Future.

There is such a tremendous amount of information, much of it difficult to understand, that people defensively tune out. Each party spins the mass of data to their advantage. Focusing on a small set of indisputable pieces of data, like this one, is critical to giving people the understanding and leverage to create actual change. Politicians would be wise to keep this in mind.

One of the measures of fiscal soundness of a country is the amount of debt it has relative to its GDP. The larger the debt, everything else being equal, the lower the level of fiscal soundness. Consider the following data for this measure across some developed countries:

Figure 4-4

Japan	220
US	92
Canada	84
France	82
Germany	80
UK	77

As can be seen Figure 4-4, the US already has the second largest amount of debt relative to the size of our economy of any of the countries listed, and the 344% shown in Figure 4-3 is off the charts. This situation has two undesirable consequences: (1) other countries may be less willing to lend us money to support the astronomic level of spending in this country, which would have a negative effect on real growth in the economy, and (2) the debt must be paid back, which presents a financial burden on the economy as well.

The massive amount of US debt currently amounts to about $50,000 for every man, woman, and child in the country. Today's children and their children will grow up with an individual financial burden which has no precedent, except perhaps at the end of Second World War. This makes it even more difficult to continue the tradition in the US of each generation having a higher standard of living than the preceding one.

Another measure of the financial soundness of a country is the size of the budget deficit compared to GDP. Figure 4-5 shows how the US compares to other developed nations using this metric:

Figure 4-5

United States	10.6%
UK	10.4
France	7.0
Canada	5.5
Australia	4.6
Germany	3.3

The US is at the top of the list in terms of budget deficit relative to GDP, and the level of our deficit is as much as twice that of some other developed countries.

Reduce Spending or Raise Taxes

It is clear from the above data that something has to be done to reduce the astronomic budget deficits and debt burden.

It is fashionable these days to talk about raising taxes on the "rich". Of course, this approach requires the monumentally arrogant assumption that government officials know what being" rich" is. Why raise taxes on the rich?

In fact, almost one-half of the American citizens currently pay no tax at all. This is the result of decades of building elaborate programs, like Social Security, Medicare, and Medicaid, to provide a cascade of benefits for those who are deemed to be unable to care for themselves. These programs were initially created to provide only a safety net, which would provide for those who had somehow fallen between the cracks of our free enterprise system.

These programs have been encouraged to grow apace, without any governor in terms of budgeting, by both politicians who get paid to deliver more and more benefits and by the beneficiaries of these programs.

These programs are, of course, paid for by those who are still paying taxes. Is it any wonder that the 50% who are not paying any taxes are campaigning for those who are still paying taxes to be forced to foot the bill for these programs? After all, it is the only way these programs can continue to grow.

There are two problems with the idea of raising taxes on the rich: (1) they already pay an extraordinary amount of the total tax bill, and (2) it is ever and always the case that when taxes are raised tax revenue goes down and the proportion paid by the rich goes down.

Figure 4-6[5] shows the percentage of total taxes paid by each of six income groups. The top 1% pay 38% of the total taxes, and the top 10% pay 70% (38% +21%+11%, the first three income categories in Figure 4-4) . Liberals talk about the rich paying their fair share. This seems like more than a fair share to me.

Figure 4-6

The Top 10 Percent of Earners Paid 70 Percent of Federal Income Taxes

Top earners are the target for new tax increases, but the U.S. tax system is already highly progressive. The top 1 percent of income earners paid 38 percent of all federal income taxes in 2008, while the bottom 50 percent paid only 3 percent. Forty-nine percent of U.S. households paid no federal income tax at all.

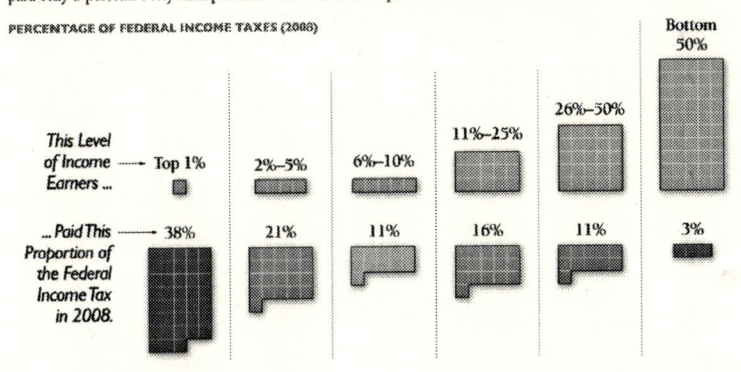

PERCENTAGE OF FEDERAL INCOME TAXES (2008)

This Level of Income Earners...	Top 1%	2%–5%	6%–10%	11%–25%	26%–50%	Bottom 50%
...Paid This Proportion of the Federal Income Tax in 2008.	38%	21%	11%	16%	11%	3%

Source: Tax Foundation and Internal Revenue Service.

Federal Revenue Chart 2 • 2011 Budget Chart Book • heritage.org

However, if you think about it, this analysis is not complete. The percentage of income taxes paid by a segment of society should be related to the percentage of income generated by that segment. The following data for 2008 are provided by the IRS:

Figure 4-7

Share of Income vs. Share of Taxes Paid

	Top 1%	Top 5%	Top 10%	Top 25%	Top 50%
% of adjusted gross income	20	35	46	67	87
% of total income taxes paid	38	59	70	86	97

You can see that while the top 10% of income producers pay 70% of the total income taxes paid, they only earn 46% of the total adjusted gross income. In fact, each of the top income categories shown pays a higher percentage of total income taxes than its share of adjusted gross income.

There is a pernicious argument from the Obama administration that the "rich" do not pay their "fair share" of taxes, which then provides a justification for government mandated redistribution of income. The concept of "fair share" is in the eye of the beholder. Because it cannot be defined, it cannot be the basis for any public policy. However, even if it were, it is hard to make the argument, based on the data in the above table, that the rich are not already paying their fair share by anyone's definition.

The worst part about this situation is that even if **all** of the income of the rich were taxed away, the resulting revenue would cover only about 20% of the current budget deficit. That is not going to happen, in the first place. However, even if it did, the heaviest burden of tax increases would still by definition fall on the middle class, because they would be forced to make up the 80% shortfall. As Willie Sutton said when asked why he robs banks, "That's where the money is."

This is confirmed by the following. "In 2009, 237,000 taxpayers reported income above $1 million, and they paid $178 billion in taxes. A mere 8,174 filers reported income above $10 million, and they paid only $54 billion in taxes. But 3.92 million reported income above $200,000 in 2009, and they paid $434 billion in taxes. To put it another way, roughly 90% of the taxpayers who would pay more under Mr. Obama's plan aren't millionaires, and 99.99% aren't billionaires."[6]

The facts here are indisputable. If the budget gap is to be closed by raising taxes, the burden will fall most heavily on the middle class.

The other reason why raising taxes on the rich is an unproductive idea is that the total revenue raised is never what the tax increasers expect it to be. Revenue is always disappointing, because the rich will do whatever they can to reduce their taxable income and therefore the taxes they pay. They can do this in a number of legitimate ways through increasing the proportion of their income which is not taxable (e.g., municipal bonds) and/or reducing their total income by reducing their investment in productive, profit-making activities.

All of this is beside the point in any case. As demonstrated in Figure 2-7[7] in Chapter 2, we do not have a revenue problem; we have a spending problem. In the Obama budget, revenue is scheduled to be 18.4% of GDP in 2021, which is in line with the long-term average of 18.0%.

How can anyone claim that we have a revenue problem when clearly revenue as a percentage of GDP is normal? The only conceivable reason for raising taxes on anyone in the face of this logic is to promote equality of condition, not equality of opportunity. This strategy should be exposed for what it is.

Equally clearly, it can be seen that we unqualifiedly have a spending problem. The long-term average for spending as a percentage of GDP is 20.3%. This has been the normal amount of support the federal government has provided to the economy.

During the period from roughly 1980 through the mid-nineties, spending was running ahead of this long-term average. Late in the decade of the 2000s, this figure approached 25%, and it is scheduled, based on the President's budget, to rise to 26.4%.

This astronomic amount of spending relative to GDP is unsustainable. However, the difficulty is that it is not easy to reduce. More and more people are beneficiaries of the spending on entitlements, which are such a large proportion of the budget, and they will give up their benefits very grudgingly.

The threat that runaway spending poses for our country is one of the worst in our history, except when we have had to defend ourselves. If this threat is not addressed immediately, the country may be set on a downward spiral from which we will not be able to recover. The most important step we can take is to make massive cuts in spending.

The politicians are finally beginning to understand the seriousness of the problem, and there are a variety of efforts underway to significantly reduce spending over time. However, one of the reasons that Standard & Poors recently lowered the nation's credit rating is that they do not have confidence that the politicians will actually succeed in addressing this problem.

The only answer, as the elections in 2010 showed, is to elect politicians who will. I have tremendous confidence in the American people to ultimately make the right decision and cut spending significantly.

Jobs and Economic Growth

Even if we are successful in dramatically cutting spending to restore it to the normal relationship versus GDP, we still have an obvious and major problem. That problem is promoting economic growth and job creation.

The Obama Keynesian model which involves massive government spending and increased control of the means of production never has created jobs, and it hasn't in this case. Economic growth from the bottom of the recession is by far the worst of any of the last four recoveries, as will be shown subsequently, job growth has been anemic most of this period, and the unemployment rate has been much higher than the administration advertised

This program has been a colossal failure. All we have to show for it is a massive increase in debt.

The tried and true approach to promote economic growth and increase employment is to cut marginal tax rates. I would like to provide some excerpts from "The Historical Lessons of Lower Tax Rates" published by the Heritage Foundation[8].

"There is a distinct pattern throughout American history: When tax rates are reduced, the economy's growth rate improves and living standards increase... history tells us that tax revenues grow and "rich" taxpayers pay more tax when marginal tax rates are slashed.

Conversely, periods of higher tax rates are associated with sub-par economic performance and stagnant tax revenues. In other words, when politicians attempt to "soak the rich", the rest of us take a bath."

There have been four major episodes of marginal tax rate reductions in the United States which illustrate these points:

Lower tax rates do not mean less tax revenue

1920s

- Tax rates were slashed dramatically during the 1920s, dropping from over 70% to less than 25%. What happened?
- Revenue rose from $719 million in 1921 to $1,164 million in 1928, an increase of 61%.
- The share of the tax burden paid by the rich (those making $50,000 and more) rose dramatically from 44% in 1921 to 78% in 1928.
- According to then-Treasury Secretary Andrew Mellon:

"The history of taxation shows that taxes which are inherently excessive are not paid. The high rates inevitably put pressure upon the taxpayer to withdraw his capital from productive business and invest it in tax-exempt securities or to find other lawful methods of avoiding the realization of taxable income.

The result is that the sources of taxation are drying up: wealth is failing to carry its share of the tax burden; and capital is being diverted into channels which yield neither revenue to the Government nor profit to the people."

The Kennedy Tax cuts

- President Kennedy realized that high tax rates were hindering the economy, and he reduced the top tax rate from more than 90% down to 70%. What happened?
- Tax revenue climbed from $94 billion is 1961 to $153 billion in 1968, an increase of 62%.

- Tax collections from those making over $50,000 increased by 57%, while tax collections from those making under $50,000 increased by 11%
- According to President Kennedy:

"Our true choice is not between tax reduction, on the one hand, and the avoidance of large Federal deficits on the other. It is increasingly clear that no matter what party is in power, so long as our national security needs keep rising an economy hampered by restrictive tax rates will never produce enough revenue to balance our budget just as it will never produce enough jobs or enough profits... In short, it is a paradoxical truth that tax rates are too high today and tax revenues are too low and the soundest way to raise the revenues in the long run is to cut the rates now."

The Reagan Tax Cuts

- Early in the 1980s, President Reagan dramatically reduced marginal income tax rates. What happened?
- The share of income taxes paid by the top 10% of earners jumped significantly, increasing from 48% in 1981 to 57% in 1988. The share of taxes paid by the top 1% of earners rose even more dramatically from 18% in 1981 to 28% in 1988.
- According to Jack Kemp, one of the architects of the Reagan tax cuts, said:

"At some point, additional taxes so discourage the activity being taxed, such as working or investing, that they yield less revenue rather than more. There are, after all, two rates that yield the same amount of revenue: high tax rates on low production, or low rates on high production."

The Bush Tax Cuts

- President Bush reduced taxes significantly in 2003. What happened?
- Economic growth rates more than doubled after 2003, the tax cuts shifted even more of the income tax burden toward the rich, and capital gains tax revenue doubled.

- According to the Heritage Foundation:

    "The [110th] Congress will be serving when the first of 77 million baby boomers receive their first Social Security checks in 2008. The subsequent avalanche of Social Security, Medicare, and Medicaid costs for these baby boomers will be the greatest economic challenges of this era.

    This should be the budgetary focus of the [110th] Congress rather than repealing Bush tax cuts or allowing them to expire. Repealing the tax cuts would not significantly increase revenues. It would, however, decrease investment, reduce work incentives, stifle entrepreneurialism, and reduce economic growth. Law makers should remember that America cannot tax itself to prosperity."

In short, the rhetorical question above – Reduce Spending or Raise Taxes – is incomplete and only partially right. The partially right part is that there is there is no question that spending has to be dramatically reduced. We cannot reach a stable financial situation without taking this step.

It is incomplete, because it completely ignores the issue of economic growth and job creation. It is partially wrong, because to do this, the proper policy position is to reduce marginal tax rates significantly, not raise them. Government spending will never create long-lasting jobs.

Think about the claim made by the Obama administration that the economy would be jump started by paying for "shovel ready" projects. This was typical Keynesian "pump priming". Let's say that you pay someone to dig a hole, expecting this payment to jump start economic activity. Then, you pay someone else to fill it up. At the end of the day, you have two people with short-term money in their pockets, and neither one of them has a real job. This kind of policy is enormously counter-productive, as the President himself subsequently admitted.

The Obama administration claimed at the outset of the massive spending program that the unemployment rate would not go above 8%. It has been above 8% from the beginning, and it came close to 10%. Does this sound like a successful program?

On the other hand, as you can see above, there has been no time since at least the 1920s when significant reductions in marginal tax rates did not produce dramatic increases in economic growth. Furthermore, if your objective is to promote class warfare and raise taxes on the most productive members of society, then the best way to "soak the rich", which some people may find counter-intuitive, is to reduce marginal tax rates.

Non-inflationary economic growth is what creates jobs and lifts all boats. That should be one of the primary objectives of any administration. Consider Figure 4-8[9], which shows the quarterly growth rate in GDP "After Deep Recessions" for two particular recoveries over the subsequent seven calendar quarters after the recession bottomed out. The two economic recoveries shown are: (1) the recovery after the Reagan tax cuts (Q4:1982 – Q2:1984) compared with (2) the recovery after the massive Obama Keynesian stimulus which pumped vast amounts of government money into the economy (Q3:2009 – Q1:2011).

Figure 4-8

QUARTERLY
GDP GROWTH RATES
AFTER DEEP RECESSIONS

1982	%	2009	%
IV	0.3	III	1.6
		IV	5.0
1983			
I	5.1	2010	
II	9.3	I	3.7
III	8.1	III	1.7
IV	8.5	III	2.6
		IV	3.1
1984			
I	8.0	2011	
II	7.1	I	1.8

The comparison is quite stark. The average quarterly growth rate for the Reagan recovery was 6.6%, while this figure for the Obama recovery was only 2.8%, well less than half the rate in the Reagan recovery. In the Reagan recovery, six of the seven quarters had growth rates in excess of 5%, while the Obama recovery had only one quarter in which the growth rate was as high as even 5%.

You can see that the Obama approach has been very costly for the people of this country, both the employed and the unemployed, in terms of economic growth and employment opportunities. This is an outrageously poor performance, and the Obama administration must be called to account for it.

Conclusion

For many decades, our country has been moving steadily toward Progressivism and away from the Founders' concept of limited government. The fact that we are demonstrably as far along in the direction of some of the concepts promoted by Marx as we are should be a wake-up call for the American people. This movement is being dramatically accelerated under the Obama administration. The concept of limited government is becoming a concept in name only.

The fundamental and dramatically effective concepts of individual liberty, free enterprise, and reduced regulatory burden are being challenged as never before in our history. We must Restore the Future through a Second American Revolution, and the time to do so is growing short.

Bottom Line

- The society created by the Founders is based on limited government
- On a spectrum with Communism on the left and limited government on the right, we are no longer anchored to the right.
- Much of the legislation in this country in the last 100 years has been designed to promote equality of condition. This governmental objective eliminates equality of opportunity, and it is a failed system with no historical success.

- Equality of opportunity is the only governmental objective which promotes liberty and which allows each person to achieve as much of their potential as they possibly can.
- 58% of President Obama's latest budget would be spent on entitlements, which is an absurdly large number, and this situation results from the failed efforts to create equality of condition.
- With respect to two key measures of a country's fiscal soundness, debt as a percentage of Gross Domestic Product (GDP) and the budget deficit as a percentage of GDP, the US is in dangerously poor condition.
- The two major alternatives currently being considered to reduce the current massive deficits are Reduce Spending or Raise Taxes. The historical data show clearly that the best alternative to reduce deficits is neither one. The only alternative to pursue, with a proven track record, is to reduce marginal tax rates.

CHAPTER 5
ON LIBERTY

"Freedom is never more than one generation away from extinction. We didn't pass it along to our children in the bloodstream. It must be fought for, protected, and handed on for them to do the same, or one day, we will spend our sunset years telling our children and our children's children what it was once like in the United States when men were free."

> President Ronald Reagan

"Liberty can be lost, and it will be, if the time ever comes when these documents (Declaration of Independence and the Constitution) are regarded not as the supreme expression of our profound belief, but merely as curiosities in glass cases"

> President Harry Truman

"Give me Liberty, or give me Death"

> Patrick Henry

Chapter 2 and Chapter 4 paint a very bleak picture. Individual liberty appears to be everywhere on the run, and the Rule of Law, which protects individual liberty and which is discussed in Chapter 6, is being eroded. The purposes of this Chapter are to define liberty, which has been the basic building block of the American experience for over 220 years, and to determine the standing among our people at the beginning of the 21st Century of this critical concept.

Over time, memories fade, education in the importance of liberty is reduced or neglected, modern life has become very complex and time-consuming, and people forget for what they should be fighting. That is why it is important to start our discussion of liberty at the beginning.

Remember what Thomas Jefferson said in the Declaration of Independence:

"We hold these truths to be self- evident, that all men are created equal, that they are endowed by their Creator with certain unalienable Rights, that among these are Life, Liberty, and pursuit of Happiness..."

The first paragraph of the Constitution reads:

WE THE PEOPLE of the United States, in Order to form a more perfect Union, establish Justice, insure domestic Tranquility, provide for the common defence, promote the general Welfare, and secure the Blessings of Liberty to ourselves and our Posterity, do ordain and establish this Constitution for the United States of America"

Our Pledge of Allegiance is as follows:

"I pledge allegiance to the flag of the United States of America, and to the republic for which it stands, one nation under God, indivisible, with liberty and justice for all."

These are all ringing endorsements of the importance of the concept of liberty. If they were on the lips of every American today, we would not be in the trouble we are in as a civilization. Unfortunately, they are not.

How many people say the Pledge of Allegiance or even know how to say it? How many people can quote portions of the Declaration of Independence? How many people, besides law students, are familiar with the philosophy and key provisions of the Constitution.

Background

It is obvious from the above and our own experiences that the term "liberty" is the cornerstone of the American experience. However, what does liberty actually mean to most people? There does not seem to be a commonly accepted definition of this concept, so that people when asked would give an answer that everyone would understand and support. This is an important part of the benign neglect described above.

The Founding Fathers were remarkably well-educated men, and they talked about liberty all the time. They were familiar with the classic writings on liberty, government, and natural rights. In addition, they seemed to have a common understanding of the concept. When "liberty" was used in the founding documents, it had a particular meaning.

I think it would be helpful to understanding how far we have departed from their vision to do two things: (1) review some of the classics, which many of them had read in the original language (Greek or Latin), to see what the Founders knew and by what they might have been influenced, and (2) determine what the concept of liberty meant to the Founders who designed a government to promote it. Then, we will be better able to see where we stand today.

The purpose of the first part of this section is not to provide a complete history of liberty – there are several good books which have already done that, including A Brief History of Liberty by Brennan and Schmidtz. The purpose is to highlight some of the key historical developments in the concept of liberty.

Philosophers and politicians have been talking about liberty for centuries, and my intention is to address some of the authors who may have had an influence on the Founders.

The Greek philosophers, whose words are remarkably well preserved thousands of years after they were spoken, present a mixed picture. I believe that this mixed picture presents key issues which were on the minds of the Founders and which, as an aside, are extremely relevant today.

Let's consider the views of Socrates and Plato, his most famous student. "Both thinkers (Socrates and Plato) demanded that statesmen should be wise. However, to Socrates, this desideratum implied that statesmen should be aware of how little they knew, while Plato felt that those with the best possible education thereby acquired the right to exercise "sophocracy", the rule of the wisest men that the rest of the population ought to be subjected to."[1]

The same author writes that "Socrates would have loved the American Constitution. After all, he observed the principle 'live free or die' and paid for his life with it. Socrates insisted on the rule of law, refusing to submit to the "rule of man"[2]. "Live free or die" echoed down the ages and showed up in the immortal words of Patrick Henry shown at the beginning of this chapter.

On the other hand, one author has described Plato as "the great enemy of liberty"... "The statist views expressed in this book (The Republic) have coloured and influenced the minds of millions of "educated" people ever since.

These "educated" people see themselves, following Plato, as being superior to others and therefore entitled to rule over them. The most dangerous idea is that of a "philosopher king" – a favorite theme of paternalistic politicians and bureaucrats who imagine that they are somehow gifted enough to tell us what to do with our lives."[3]

Here is more about Plato from the same author. "A central thesis of The Republic is that personal freedom and political liberty are not important. According to Plato, it is far more necessary to improve the *moral* condition of a society than it is to protect the *political rights* of its citizens."[4]

"For Plato, the purpose of the state is to improve the condition of the state itself. The condition of those who ultimately comprise the state – its citizens – is largely irrelevant. But for a liberal, when human rights are sacrificed on the altar of something supposedly more important than those rights themselves the result is inevitably tyranny."[5]

Plato's views were emphatically rejected by the Founders. For them, Plato had things exactly backwards. However, it is interesting to note that, despite this rejection, his words are the foundation of modern-day Progressivism.

These were not just philosophical arguments. Some of the ideas of the philosophers were actually put into practice in Greece, Rome, and other city-states. The best example, perhaps, is the experiment in ancient times with pure democracy. James Madison, writing in <u>Federalist #10</u> said the following: "...it may be concluded that a pure democracy, by which I mean a society consisting of a small number of citizens, who assemble and administer the government in person, can admit of no cure for the mischiefs of faction." What he means is that it is easy in a pure democracy to have "tyranny of the majority", because there is nothing to protect the interest of the minority from being overrun by the passion of the majority.

Further on in <u>Federalist #10</u>, Madison continues, "Hence it is that such democracies have ever been spectacles of turbulence and contention: have ever been found incompatible with personal security or the rights of property; and have in general been as short in their lives as they have been violent in their deaths.

Theoretic politicians, who have patronized this species of government, have erroneously supposed that by reducing mankind to a perfect equality in their political rights, they would, at the same time, be perfectly equalized and assimilated in their possessions, their opinions, and their passions.

The two great points of difference between a democracy and a republic are: first, the delegation of the government, in the latter to a small number of citizens elected by the rest; secondly, the greater number of citizens, and greater sphere of country, over which the latter may be extended."

There were a number of significant events in the Dark Ages and the Middle Ages, such as the signing of the Magna Carta in 1215 and Martin Luther's posting of his ninety-five theses in 1517, which was the beginning of the Protestant Reformation. However, this was not a period when the philosophical thinking about liberty was advanced dramatically. That ultimately happened in the Age of Enlightenment.

From the standpoint of liberty, the signing of the Magna Carta was the signal event during this period.

"The Magna Carta is a document that King John of England was forced into signing. King John was forced into signing the charter because it greatly reduced the power he held as the King of England and allowed for the formation of a powerful parliament. The Magna Carta is considered the founding document of English liberties and hence American liberties. The influence of the Magna Carta can be seen in the United States Constitution and the Bill of Rights."[6]

There was an explosion of interest in and writings about liberty in the late 17th Century, prior to the American Revolution. Two of the most important writers and thinkers of this period were Algernon Sydney and John Locke. In fact, "Thomas Jefferson believed Sidney and Locke to be the two primary sources for the Founding Fathers' view of liberty."[7]

"Sydney is an English martyr for republican government [and a pioneer in natural rights theory]. He was executed in 1683 for conspiring to kill King Charles II, and the only evidence against him was an unpublished draft of the "Discourses on Government". "Sydney [was] credited with authoring the original constitution for the Carolina colony and assisting his good friend William Penn in drafting the original Constitution of Pennsylvania in 1682."[8]

Furthermore, there is some speculation that he was involved with New York's Charter of Liberties and Privileges. "The 1683 Charter (constitution), [which was ratified shortly before his execution], was the first of its kind in America, and all of our constitutions retain essentially the same form. It was the first constitution to have the phrase 'the people' in it. Thomas Jefferson ... would later draw heavily on Sydney's Discourses in drafting the Declaration of Independence."[9]

Sydney was one of the first to call for revolution under certain circumstances and for government by the people. He said, "That which is not just, is not law: and that which is not law, ought not to be obeyed."[10] He was violently opposed to the monarchy, and it is clear that he felt that the divine right of kings was not the way to determine what is just.

Sydney went on to say, "I am persuaded to believe that God had left nations to the liberty of setting up such governments as best pleased themselves, and magistrates were set up to the good of the nations, not nations for the honor and glory of the magistrates".[11] Although this was said a hundred years before the Constitutional Convention, this quotation was very similar to the thinking of the Founders.

In fact, as founder of the University of Virginia, Thomas Jefferson issued this statement; "Resolved, that it is the opinion of this Board that as to the general principles of liberty and the rights of man, in nature and in society, the doctrines of Locke, in his 'Second Treatise on Government concerning the true original extent and end of civil government', and of Sydney in his 'Discourses on Government', may be considered as those generally approved by our fellow citizens of this, and the United States."[12]

"John Locke was an English philosopher and physician regarded as one of the most influential of Enlightenment thinkers." "Such was Locke's influence that Thomas Jefferson wrote: "Bacon, Locke, and Newton... I consider them as the three greatest men that have ever lived, without any exception, and as having laid the foundation of those superstructures which have been raised in the Physical and Moral sciences."[13]

"Locke's political theory was founded on social contract theory. Like Hobbes, Locke assumed that the sole right to defend in the state of nature was not enough, so people established a civil society to resolve conflicts in a civil way with help from government in a state of society." "Locke also advocated governmental separation of powers and believed that revolution is not only a right but an obligation in certain circumstances."[14]

In an unprecedented proceeding – the Constitutional Convention – and in the space of only four months in the summer of 1787, the Founders created the Constitution, based on natural rights, limited federal government, separation of powers and government of the people, by the people, and for the people.

This was the greatest enshrinement of liberty in the history of the world. It is interesting to note that intellectual focus on the philosophy of government in the more than 220 years since the founding of the republic has produced no major advances anywhere in the world. Other kinds of government have been tried, and all of them have failed. That is why the American system of government is still a beacon for people all over the world.

Here are some interesting reflections by the founding generation about what they had created, together with the identification of some of the things which they warned could go wrong:

(1) "They that can give up essential liberty to obtain a little temporary safety deserve neither liberty nor safety"[15] - Benjamin Franklin

  The point is that liberty must be fought for every day, as President Reagan said in the quotation above.

(2) 'If we are to be directed from Washington when to sow and when to reap, we should soon want bread"[16] – Thomas Jefferson

  This, of course, is the argument for limited government. It was inconceivable to the Founders that centralized government planning would work, because no collection of wise men will ever to be able to make all the correct decisions necessary to promote liberty for its citizens.

(3) "The natural progress of things is for liberty to yield and government to gain ground."[17] – Thomas Jefferson

This is a very prophetic comment, given what has been happening in this country over the last 100 years. This is why it is so difficult to roll back the achievements of the Progressives.

(4) "The democracy will cease to exist when you take away from those who are willing to work and give to those who would not"[18] – Thomas Jefferson

This is another very prophetic comment. It anticipates exactly the situation in which we find ourselves today, when almost 50% of the people in this country do not pay any income taxes to support the government from which they benefit.

(5) "Power over a man's substance is power over his will"[19] – Alexander Hamilton

Hamilton seems to be saying that giving the government the ability to take what it wants from you gives the government the ability to reduce your liberty.

(6) "Government... should be formed to secure and enlarge the exercise of the natural rights of its members: and every government which has not this in view as its principal object is not a government of the legitimate kind."[20] – James Wilson

James Wilson was one of only six of the Founders who signed both the Declaration of Independence and the Constitution. The point he makes here is critically important. What he is saying is that government should be created for the express purpose of maximizing the liberty of its citizens, and he certainly would be speaking for the other Founders as well. This view will be an important part of the subsequent discussion.

This is certainly not the view of the Progressives. They promote a government of the elite who know better what the people want and should have than the people themselves. Those who support maximizing liberty have faith in the people: the Progressives have faith only in themselves.

As you can see, not only did the Founders create a form of government which provided maximum liberty for its citizens to be the best that they could be, which had no precedent before they did it, they warned us how it could be brought down. They told us for what to watch out, and it is the realization of these warnings and others which contributes to the call in this book for a Second American Revolution to Restore the Future.

The Founders had built the new country on the strongest possible foundation, based on the best thoughts in the world at that time on liberty, form of government, individual rights, and the origin of those rights, and they had the foresight to provide a way to amend what they had created if better ideas came along.

If you would like to understand their reasoning and the alternatives which they considered, based on their knowledge of world history to that point, I suggest that you read the Federalist Papers, a collection of 85 essays written in support of the Constitution by James Madison, Alexander Hamilton, and John Jay.

"One of the most common misconceptions in the United States is that people's rights come from the Constitution. Without the Constitution, it is believed, people wouldn't have such rights as freedom of expression and religion. People should be grateful to the founding Fathers, it is said, for establishing the vehicle by which people could have such rights as life, liberty, and property... Nothing could be further from the truth."[21]

This quotation from the Future of Freedom Foundation suggests that these rights are antecedent to the Constitution. Frederic Bastiat agrees. Bastiat, who was a French classical liberal theorist and political economist, hit the nail on the head when he said in about 1850 that, "Life, liberty, and property do not exist because men have made laws. On the contrary, it was the fact that life, liberty, and property existed beforehand that caused men to make laws in the first place."[22] Thus, the very essence of the Declaration of Independence and the Constitution is confirmed once again well after their creation.

The founding documents confirmed what already existed, namely natural rights. The remarkable achievement of the Founders was to set up a government which was designed to protect and promote them.

Definition of Liberty

As we have seen, some very historically well-grounded and visionary men have talked about liberty for centuries. However, there is one thing which continues to elude us, and that is a definition of liberty. Everyone seems to know what it is, but a good definition is necessary, so that everyone will have a common conception of what it is for which we are fighting.

The best discussion of the definition of liberty is found in an excellent book written by Matthew Spaulding – We Still Hold These Truths. The following quotations are from this book.

"So important is the concept [of liberty] that English - unlike any other language – has two words to describe it: *liberty* as well as *freedom*. We tend to use the term *freedom* more nowadays, for it has a powerful and evocative ring to it. The Founders preferred and widely used the word *liberty*."[23]

Spaulding continues, "There is a difference between these two terms that helps us understand the Founders' concept of the principle. Freedom is understood to be more expansive, and suggests a general lack of restraint..., as we speak of the United States as a 'free society'. It is often used to suggest a more open-ended sense of autonomy, meaning that we are free to do whatever we want."[24]

The Founders had a different view, and it is important to understand it, because this view is the basis for our entire society. Spaulding says, "...freedom must be understood within the context of constitutional and moral order, which means reasonable limits and cultural bounds." Liberty means the rightful exercise of freedom, the balancing of rights and responsibilities."[25]

To help us understand the difference between liberty and freedom more clearly, Spaulding continues, "All animals can be said to have freedom. Men can be free, but so can fish in the ocean or birds in the sky. But liberty is an inherently human word. While we say that man has liberty or is at liberty to do something, we do not say the same of animals, because animals lack a rational capacity to choose their own actions."[26]

"In the American tradition, liberty was never understood to mean anything and everything, but came with duties and obligations appropriate for human self-government."[27]

When people figuratively sign a social contract, they give up freedom to gain liberty. This highlights in another way the difference between the two words. It would make no logical sense to give up freedom to get exactly the same freedom inside a social contract arrangement. Freedom is freedom, and in this case there would obviously be no need for a social contract.

Therefore, what you get with a social contract is different than simply freedom. What you get, as it turns out, is the personal liberty to be what you want to be, with the only restraints being those which come with the social contract.

In other words, freedom is the ability to do whatever you want whenever you want. Freedom is theoretically the greatest in the state of nature described by Rousseau. On the other hand, liberty is freedom to do whatever you want whenever you want, but only within the context provided by the social contract.

Suppose that you have an opportunity to sign two different social contracts – one with a society which is organized to maximize liberty and one which allows liberty to be compromised for the best interests of the society as a whole by the elites, as in the Progressivist view. Obviously, most people would choose the first option, because they get the most liberty for the freedom which they give up by becoming a part of the society in the first place.

Clearly, then, the first society when it competes with a society such as the second one indicated above would have an enormous competitive advantage in maintaining a stable society by maximizing liberty. This was the situation with the Cold War, and President Reagan was right to believe that the Soviet Union was unstable because of its lack of commitment to liberty.

Since the beginning of the country, we have celebrated American exceptionalism. Our feeling about our country is not meant to imply a criticism of any other country or form of government. It may be that other countries consider themselves to be exceptional as well. We have no monopoly on what we have created, and any other country can follow in our footsteps whenever they choose. A number of countries today are trying to do just that.

What we mean by American exceptionalism is best described in the words of the Founders:

- "The establishment of our new government seemed to be the last great experiment for promoting human happiness by reasonable compact in civil society."[28] – George Washington
- 'Is it not the glory of the people of America, that, whilst they have paid a decent regard to the opinions of the former times and other nations, they have not suffered a blind veneration for antiquity, for custom, or for names, to overrule the suggestions of their own good sense, the knowledge of their own situation, and the lessons of their own experience?"[29] – James Madison
- "They accomplished a revolution which has no parallel in the annals of human society. They reared the fabrics of government which have no model on the face of the globe."[30] – James Madison

These words of the Founders are as true today as they were when they were written. We are exceptional because of the government which was created by the

Founders. There can be no doubt that this government has provided more individual liberty, as defined above, than any other in the history of the world. Since that is one of the key responsibilities of government, by definition we are exceptional.

Liberty Today

Now that we have reviewed the evolution of the concept of liberty, reviewed the thinking of some of the great men who contributed to the evolution, and developed a definition of liberty in the first place, we now have a context within which to evaluate the condition of liberty in the situation in which we find ourselves at the beginning of the 21st Century.

There are two principal criteria for this evaluation:

- The principal responsibility of a government in a society in which people voluntarily sign a "social contract" is to maintain and promote liberty for these people, and
- Liberty is defined as 'the rightful exercise of freedom, the balancing of rights and responsibilities".

Keep in mind that all the evidence indicates that government and liberty are a "zero-sum" game. What this means is that there cannot be both an increase in government involvement and an increase in the liberty at the same time. If you want more of one, you have to give up some of the other.

The more government involvement in the society, the less liberty there is and vice versa. The powers of the federal government are strictly limited by the Constitution to provide the maximum amount of liberty for American citizens. There can't be any more liberty than this, but there can be less.

Now let's consider a list of current activities of government which I believe have the effect of increasing the involvement of government in our society and decreasing liberty. Use the two criteria outlined above to draw your own conclusions about whether or not you agree.

- Economic

    - Promoting a massive Keynesian government spending program to deal with the recession, which had the following effects: an unprecedented increase in the national debt, massive deficits, an unbelievable financial burden for our children and grandchildren, the slowest rate of overall economic recovery from a recession in modern history, and the worst period of job recovery after a recession since the Great Depression
    - Tolerating a situation in which half the people in this country do not pay income taxes. This is an example of how you might reach a tipping point. How do you get those who currently do not pay taxes to pay them? The only and obvious answer is to reduce marginal tax rates and broaden the tax base.
    - Engaging in class warfare by deciding in the most arbitrary manner at what point you make "too much" and should pay even more in taxes. This amounts to robbing Peter to pay Paul, when Peter is the most successful and job-generating part of the economy and Paul is just the opposite
    - Maintaining a corporate tax rate which is one of the highest in the developed world. This encourages businesses to relocate to offshore environments in which the tax environment is much more friendly, and this, in turn, reduces jobs in this country.
    - Arranging through appointments to the National Labor Relations Board to dictate that Boeing cannot move some production from the state of Washington to the state of South Carolina, which is a "Right-to-Work" state. A right-to work state does not require employees to be members of unions. The rationale of the NLRB has to be that Boeing's move reduces union employment in the economy as a whole without regard for the negative impacts on efficiency and global competitiveness such a requirement would create.
    - Creating an unknown number of "czars", who are not subject to approval by Congress, who control things like the magnitude of wage and salary increases in the industries for which they are responsible,

whether certain companies should survive, which contractors doing business with the government must hire union labor, and who should benefit and who should be hurt in bankruptcy
- Not promoting free trade, either because other countries with whom we might have trade agreements do not have the labor laws which we have in this country or because there is concern about job losses in this country. As will be demonstrated in Chapter 9, free trade always creates net new jobs, and a policy of not promoting free trade is immensely counter-productive.
- Creating such a labyrinth of regulations in such areas as the environment, taxes, human resource management, and product liability that businesses are paralyzed into inactivity, because the environment is so uncertain. This is one of the reasons for the unprecedentedly slow economic recovery.
- Promoting an anti-business environment. Politicians seem to spend most of their time identifying and trying to fix what's" wrong" with business, instead of identifying and implementing ways to promote businesses and help them grow. This is obviously counter-productive, because businesses, particularly small businesses, create the jobs in this country. Governments don't create real jobs, except in the public sector which they control through spending increases.

- Social

  - Promoting equality of condition instead of equality of opportunity. Maximum growth and prosperity for everyone are created by maximizing liberty. Equality of condition imposes a grossly inefficient straightjacket by forcing business to use employees, one selected from column A (some ethnic group, for example), column B (different group), etc., who are not qualified.
  - Enacting a nationalized healthcare program which requires people to buy health care or pay a fine and which inserts the government between patients and doctors. Programs like this have been operating in the UK and Canada for some time, and they have

had disastrous results in terms of cost and patient care. This is a very clear example of how liberty is being reduced by increasing government involvement.
- Encouraging "tyranny of the minority". If only one person has an objection, an entire community can be barred from doing something that all of the other people in the community find acceptable.
- Encouraging Balkanization of the population, so that it is acceptable to be a hyphenated American – black-American, Asian-America, etc. A policy like this promotes inefficiency and discord. Over our entire history, we have encouraged immigrants to come here legally and assimilate into the American system. What we now tolerate is illegal aliens coming to this country and doing their best not to assimilate.

- **Other**

  - Having no immigration policy, and encouraging illegal aliens to stay in this country and benefit from our system – sending their children to public schools, getting health care, etc.- with their expenses paid for by citizens who are here legally. This is another example of robbing Peter to pay Paul.
  - Apologizing for those in this country who believe in American exceptionalism and denying that such a thing exists in the first place.
  - Accepting an educational system which produces students whose test scores are no better than mediocre when compared to those of students in other countries.
  - Reducing defense spending as a percentage of GDP to an historically low level which is not consistent with the number and variety of challenges which we face around the world.

Unfortunately, this is not a complete list, and readers can create and evaluate their own list of activities. However, it is more than sufficient to demonstrate how serious the problems are and why individual liberty is more threatened today than it has ever been in the history of the Republic.

At this point, it should be clear that one of the key driving forces behind the Second American Revolution has to be the restoration of limited government and the maximization of personal liberty.

What we are talking about is really not very complicated. The following words of Lord Acton, a British historian and political philosopher writing in the mid-nineteenth century, should be on the lips of every American and particularly those who run for public office:

"Liberty is not a means to a higher political end, it is itself the highest political end."[31]

It was not very long ago when a leading politician actually recognized the importance of liberty and asked the American people to rise to the challenge of defending it. Remember the immortal words of President John Kennedy at his inauguration 50 years ago in January, 1961:

"Let every nation know, whether it wishes us well or ill, that we shall pay any price, bear any burden, meet any hardship, support any friend, oppose any foe in order to assure the survival and the success of liberty [emphasis added]."

What we have concluded so far is that people voluntarily participate in a society with an implied social contract and that the people create a government within the social contract which has as one of its two objectives maximizing the liberty of its citizens. How is a potentially unstable situation controlled so that liberty is protected from the government and other citizens? This is the subject of the next chapter.

Bottom Line

- Liberty is the cornerstone of the American experience
- There are significant differences between a democracy and a republic
- Algernon Sydney and John Locke provided much of the philosophical foundation for the Declaration of Independence

- There have been no advances in the concept of government anywhere in the world over more than 220 years which provide as much liberty as the American system of government.
- Liberty and freedom are different concepts. Liberty is freedom within a social contract.
- The less government involvement in the society the more liberty exists and vice versa.
- American exceptionalism is real, and it is, by definition, derived from our form of government.
- There is a long list of examples of the present administration acting to reduce liberty.

CHAPTER 6
THE RULE OF LAW

The Founders understood that there are two fundamental ways in which government can exercise its authority. The first is a system of arbitrary rule, where the government decides how to act on an ad hoc basis, leaving decisions up to the whim of whatever official or officials happen to be in charge; the second way is to implement a system grounded in the rule of law, where legal rules are made in advance and published, binding both government and citizens and allowing the latter to know exactly what they have to do or not do in order to avoid the coercive authority of the former.

> Ronald Pestritto
> Hillsdale College Associate Professor

The rule of law is a legal maxim that provides that no person is above the law, that no one can be punished by the state except for a breach of the law, and that no one can be convicted of breaching the law except in the manner set forth by the law itself. The rule of law stands in contrast to the idea that the leader is above the law, a feature of Roman law, Nazi law, and certain other legal systems.

> Wickipedia

The Constitution is our fundamental law because it represents the settled and deliberate will of the people, against which the actions of government officials must be squared. In the end, the continued success and viability of our democratic Republic depend on our fidelity to, and the faithful exposition and interpretation of, this Constitution, our great charter of liberty.

> Edwin Meese
> Former US Attorney General

As mentioned in Chapter 1, the government of any society has to operate with a system of laws which govern how it functions. If and as its Rule of Law is compromised or vitiated, the society necessarily starts falling apart, and it eventually declines into chaos.

The system of laws for our country starts with the Constitution and the Bill of Rights. The purposes of this chapter are to: (1) describe the origin of our Rule of Law, (2) analyze the fidelity to our Rule of Law in our country today, and (3) to highlight those areas which represent major departures. The magnitude and the frequency of any departures will tell us about the extent to which this country has begun an inevitable decline.

On the other hand, this same identification will allow us to see what needs to be done to restore the Rule of Law to its rightful place in society, so that any decline is reversed. As is the case with the promotion of liberty, the promotion of the rule of law is an important part of the Second American Revolution to Restore the Future.

Background

As was the case with liberty in the last chapter, it is necessary to develop a definition of the rule of law, so that everyone knows what we are talking about. What is the meaning of the rule of law, and where does the rule of law fit in with the concept of a social contract society in which government has the primary responsibility of promoting liberty?

It is obviously necessary in any society, as discussed earlier, to have rules of the game. People need to know how to live their lives as part of the society. Throughout the long sweep of history, until fairly recently, these rules were created and enforced by force of arms. Whoever had the power, whether a tyrant, military conqueror, or monarch, made up the rules. No one had any interest in asking the people who lived in these societies what they thought.

Gradually, however, the idea of a contract between ruler and ruled about certain kinds of issues began to develop in the western world. The best example of this is the signing of the Magna Carta by King John in 1215. "In its famous thirty-ninth clause, He promised that "[n]o free man shall be taken, imprisoned, disseized, outlawed, or banished, or in any way destroyed, nor will he proceed against or prosecute him, except by the lawful judgment of his peers and the Law of the Land."[1]

This is an extraordinary development which had no precedent. The King was forced by the barons to agree that the "Law of the Land" was superior to human rulers. This concept was confirmed over 400 years later. "The ultimate outcome of the Glorious Revolution of 1688 in England was permanently to establish that the king was subject to the law."[2]

These kinds of developments echoed down the centuries, and they eventually led to a concept of common law in England. The English do not now have and never did have a constitution like America has, and the English common law, a collection of individual laws which have been enacted over the years, now governs the country.

"The classic American expression of the idea comes from the pen of John Adams when he wrote the Massachusetts Constitution in 1780, in which the powers of the commonwealth are divided in the document 'to the end it may be a government of laws, not of men."[3] This concept that laws apply to everyone was a major advance in the development of the social contract society, because it holds out the possibility of an enormous increase in personal liberty.

Naturally, the English common law, which consists of a collection of individual laws which were passed at different times, was transferred over time to the American colonies. The English common law has proven to be very effective without a constitution, but the Founders insisted that there be a written document. They knew that the American people would not accept such a key document unless everything involved was written down in clear and concise language, so that the opportunity for misinterpretation then and in the future was limited.

Therefore, the rule of law in this country, by definition, starts with the Constitution.

In <u>We Still Hold These Truths</u>, author Peter Spaulding summarized the meaning of the rule of law at the time of the Constitutional Convention. "Over time, the rule of law had come to be associated with four key components:

- First, the rule of law means a formal, regular process of law enforcement and adjudication.

    This is the idea of due process. This is a system of laws and a process for their enforcement, instead of the arbitrary will of some authority figure. Enforcement is provided by courts of law and a governmentally based law enforcement process. For such an approach to be credible, it has to be clear to the people who are governed by the laws that all citizens are treated equally.

- Second, the rule of law means that these rules are binding on rulers and the ruled alike.

    In <u>Federalist 57</u>, Madison said, "if the American people shall ever be so far debased as to tolerate a law not obligatory on the legislature, as well as on the people, the people will be prepared to tolerate anything but liberty". This quotation puts into perspective the modern-day practice of Congress to either: (1) pass laws which give them privileges which are not available to the citizens as a whole, and (2) pass laws which apply only to the citizens and not to them.

- Third, the rule of law implies that there are certain unwritten rules or generally understood standards to which specific laws and lawmaking must conform.

    These unwritten rules are not laws per se, but they are implicit in the law itself. There are a number of examples which show up in the Constitution. To illustrate the point, Madison said in <u>Federalist 44</u>, "Bills of attainder (laws which punish individuals or groups without a judicial trial), ex-post facto laws (which are laws created after an action has been taken to make that action a crime), and laws impairing the obligation of contracts are contrary to the first principles of the social compact, and to every principle of sound legislation".

- Lastly, even though much of its operation is the work of courts and judges, the rule of law ultimately is based on, and emphasizes the centrality of, lawmaking."[4]

Spaulding explains, "The rule of law – especially in terms of key procedural and constitutional concepts – stands above government... The more authoritative or fundamental laws have an enduring nature. They do not change day to day or by the whim of the moment, and cannot be altered by ordinary acts of government."[5]

The Magna Carta's phrase "Law of the Land" is written into all eight of the early American state constitutions, and it is reflected in the supremacy clause of the US Constitution. The major holders of US office take an oath to the Constitution and the laws, not to some person or governmental entity.

It is important to realize that for the concept of a rule of law to function in its necessary role to protect liberty in a society it has to be subordinate to the people. "There can be no more dangerous doctrine in a state, than to admit that the legislative power has a right to alter the constitution," wrote one pamphlet writer in 1776 under the pen name of Demophilus. "For as the constitution limits the authority of the legislature, if the legislature can alter the constitution, they can give themselves what bounds they please."[6]

Put another way, "... that Constitution must be above ordinary legislation and the changing actions of government. That is what they meant by a 'constitution', and that, by definition, meant a framework of limited government."[7]

Think about the application of the supremacy of the Constitution compared to legislation in terms of the debate about Obamacare. What the Congress under Obama has done is to restructure one-sixth of the US economy without giving serious consideration to the constitutionality of what they were doing. The point is that the Constitution is the ultimate authority, which is why there is such a vigorous debate about the constitutionality of the new law at this point. Happily, this law, by definition, cannot be put into practice without the ultimate sanction of the Constitution and the Supreme Court.

Thomas Paine, speaking for all Americans, said, "...an unwritten constitution is not a constitution at all." With this in mind, the Second Continental Congress, after the attacks at Lexington and Concord, asked the colonies to develop their own state constitutions to include the right to free speech, the right to bear arms, and responsibilities which each state thought appropriate. It is interesting to note that, "the oldest written constitution in the world is the one John Adams wrote for Massachusetts in 1780"[8].

As we know, the Founders created our Constitution in the remarkably short period of four months in the summer of 1787. Winston Churchill said, in commenting about the role of the RAF in the Battle of Britain, "never in the history of mankind have so many owed so much to so few". This quotation is remarkably apt to describe the American Founders and the Constitution they created. Thomas Jefferson concluded that, "the resulting Constitution is unquestionably the wisest ever yet presented to men". It still is today.

"While often overlooked, and taken for granted, the Constitution is central to American life. Not simply an organizational structure having to do with narrow legal or governmental matters, it is the arrangement that formally constitutes the American people. It orders our politics, defines our nation, and protects our citizens as a free people."[9] There can be no question about the primacy of the Constitution as the basis for our rule of law.

Opponents of the Constitution were concerned that it did not specifically provide for the rights of the citizens, except in a limited way. Addressing this concern was the unfinished business of the Constitutional Convention. In 1789, Congress created a Bill of Rights, drafted principally by James Madison, and the identified rights were added to the Constitution as the first ten amendments.

"The purpose of the Bill of Rights – stated by both the Federalists [those supporting a strong central government] and the Anti-Federalists - was to limit the federal government, not the states"[10]. It is very important to note at this point that what are specified in the Bill of Rights are what can be described as "negative liberties" (see discussion below). Free speech, the right to bear arms, right to a speedy trial, trial by jury, and freedom from unreasonable searches and seizures are specified as rights the government cannot take away. They are critically important rights which cannot be "negated".

The Founders were not naïve enough to think that what they created would last for all time exactly as it was written. Consequently, they included a very carefully designed and formal process for modifying the Constitution as time went by. To insure that the process was deliberative, reasonable, and legitimate in terms of reflecting the will of the people, the Founders intentionally made it difficult to amend the Constitution.

A proposed amendment can only be forwarded to the states for their consideration if two thirds of both houses of Congress approve it. It is then necessary for the amendment to be approved by three quarters of the states. These are formidable hurdles. "Although more than five thousand bills proposing to amend the Constitution have been introduced in Congress since 1789, there have been only seventeen additional amendments to the Constitution besides the Bill of Rights."[11]

It would appear that, based on the data, the Founders were successful in their determination to preserve the essence of what they had created and to make it difficult for amendments to be enacted. However, this is more of a problem than appears at first glance. Because it is appropriate to have an amendment process which makes amendments difficult, people have tried to end-run this process by acting to amend the Constitution by either legislative or judicial fiat.

They do this by defining rights which don't exist in the Constitution (e.g., abortion), governmental powers which don't exist (enormous expansion of the Commerce Clause), and judicial decisions which are not entirely based on interpretations of the US Constitution (considering laws from other countries).

Alexander Hamilton said in Federalist #78 that "It is far more rational to suppose that the courts were designed to be an intermediate body between the people and the legislature, in order, among other things, to keep the latter within the limits assigned to their authority. The interpretation of the laws is the proper and peculiar province of the courts... or in other words, the constitution ought to be preferred to the statute, the intention of the people to the intention of their agents".

"By cultivating and allowing the deliberative, popular will to assert, by constitutional means, its sovereign authority over government, the amending process affirms the rule of law and links our highest law back to the republican idea that government ultimately derives its just powers and legitimate authority from the consent of the governed..."[12]

What this means is that, since our Law of the Land starts with the Constitution, the only amendments to it which can be tolerated are those which are formally approved using the process the Founders created. Anything else is, by definition, a violation of our Rule of Law.

"The basis of our political systems is the right of the people to make and to alter their Constitutions of Government," George Washington wrote in his Farewell address of 1796. "but the Constitution which at any time exists, he reminded us, till changed by an explicit and authentic act of the whole People, is sacredly obligatory upon all."

"The rule of law may be the most significant and influential accomplishment of Western constitutional thinking"[13]

Current Situation

You would think that, based on the facts that: (1) the Constitution was the greatest document in the history of the world when it was created and, despite a variety of efforts to create alternative forms of government throughout the world, no one has come close to a better design, (2) the system of government which was created based on the Constitution has produced one of the most successful civilizations which has ever existed, and (3) a Rule of Law has been developed which is well-understood and accepted, the American people would be remarkably well-informed about the Constitution and very vigorous in their support for it.

Unfortunately, this is not the case. The American people are remarkably uninformed about the Constitution, because the education they receive in civics is very limited. Many of us can remember our own education and what a significant amount of it was based on civics.

Not only are our students doing poorly relative to students in other parts of the world in general (see Chapter 8), but they are doing poorly relative to what they should be learning to be vigilant about the preservation of our civilization. Remember Franklin's warning, "if you can keep it". The only way you can keep it is to be well-informed and constitutionally active at every turn.

The attitude of the American people at this point is one of benign neglect relative to the Constitution. This is intolerable, and it is part of the justification for the Second American Revolution.

There are enumerable, specific challenges to the Constitution in our country today. Some of them result from activists trying to push some agenda, and they are rarely challenged to prove the constitutionality of what they want to do. So a creeping reduction in the commitment to the Constitution is inevitable. It is critical to have a constitutionally-based dialogue about everything we do as a matter of routine.

Some of these challenges come from attempts by the executive branch to circumvent the legislative branch. "Following the November elections, when President Obama's party lost control of the House, Obama told America that 'where he can't legislate, he will regulate' And that seems to be this Administration's *modus operandi*: If Congress refuses to abide by Obama's agenda, the President's bureaucratic machine will make its own laws."[14]

"The Obama administration generally employs one of two strategies to legislate without – and often in spite of – congressional action: (1) administrative decree establishing a federal rule, or (2) a refusal to enforce existing federal law. In five separate policy areas, the President and the federal agencies under his command have spurned congressional authority to achieve Obama's objectives:

- Environmental Regulation – imposing economy-killing environmental regulations and classifying carbon dioxide as a pollutant which can be regulated by the EPA
- Labor Law – allowing the NLRB to attack Boeing's decision to re-locate a plant from the State of Washington to South Carolina - a "right-to-work" state - and drastically reducing the notice time required for votes on authorizing a union.
- Immigration Law – encouraging "prosecutorial discretion" in the treatment of illegal immigrants
- Selective Enforcement of Federal Law – stopping enforcement of laws against marijuana use and laws such as the Defense of Marriage Act, because the administration doesn't believe in them
- Regulating the Internet – allowing the FCC to override an appeals court ruling and pass the first ever regulations on internet traffic."[18]

Some of them come from regulations, which have the direct effect of reducing liberty. Did you know that, "In fiscal year 2009, new regulations costing more the $13 billion per year were adopted by the Bush and Obama Administrations, the highest annual total since 1992...The effects of such a regulatory tsunami could be disastrous – destroying jobs, threatening enterprises, and deterring new investment."[16]

This situation may be an important factor behind what is the slowest and most anemic economic recovery from a recession in the post-war period. Businesses are choking on red tape and constantly uncertain about what they can and cannot do.

"According to a 2005 study by the Small Business Administration, the total burden of federal regulation [not counting state regulation] is some $1.1 trillion

– more than Americans pay in income taxes each year."[17] Every time you create a huge new body of regulation, you have to, of course, create the vast bureaucracies which hire regulators to provide the regulatory oversight. Obama's budget for fiscal year 2011 calls for expenditures for regulatory agencies of over $59 billion.

"The Code of Federal Regulations, a compendium of all existing federal rules hit a record high of 163,333 pages in 2009, an increase of some 22,000 since the beginning of the decade."[18] What follows from this level of regulation is that businesses have to hire lawyers and other experts at great expense just to navigate through the regulatory jungle. This is the collateral damage of too much regulation.

Another example of government intrusion into society is overcriminalization. Did you know that, "the number of criminal offenses in the U.S. code increased from 3,000 in the early 1980s... to 4,450 in 2008"[19], an increase of almost 50%. Is it any wonder that there are more people in US jails per capita than in any other major country in the world?

"By consistently creating new criminal laws without consulting the special expertise of the two judiciary committees [House and Senate Judiciary Committees], Congress is endangering civil liberties and placing all Americans at risk of unjust criminal conviction for violating crimes they did not even know they were committing."[20]

Finally, some of them come from the willingness of the people to allow the development of an inordinately litigious society. Did you know that, "From 1996 through 2005, more than 135 million civil lawsuits were filed in state courts – an average of 52,000 new cases every business day – and payouts for tort losses and insurance premiums increased by 60% in inflation-adjusted dollars"[21]

"The current system...encourages plaintiffs' lawyers and activists to manipulate the civil justice system for their own ends at the expense of the public."[22] Did you know that only 15 cents of every dollar award in tort liability cases is returned to plaintiffs? The vast majority of the awards goes into the pockets of trial lawyers, who represent a powerful lobby to maintain the status quo. Did you know that the American tort system is the most expensive in the world?

This is just a small list of specific issues which represent challenges to the Constitution and to personal liberty. However, there are more important philosophical challenges to the Constitution of which most people are not aware. These probably represent the most significant challenges to the Constitution in the history of the Republic.

One of the best presentations of the challenges to the Constitution which will develop over the next few years is provided in a speech entitled "The Coming Constitutional Debate"[23] by Stephen Markman, a Justice of the Michigan Supreme Court, at an event sponsored by Hillsdale College.

All of these challenges represent attempts to amend the Constitution in ways which have nothing to do with the formal process for amending the Constitution provided by the Founders in that very document. The modern Progressives are mounting a challenge which is the most audacious and dangerous since the days of the New Deal. They are doing this, because they know that they could not accomplish what they want to accomplish with the explicit and transparent amendment process. The American people would not stand for it.

He identifies six issues which will be an important part of the upcoming "Constitutional Debate":

- Privileges or Immunities Clause

    "Since shortly after the Civil War, the privileges or immunities clause of the 14th amendment has been understood as protecting a relatively limited array of rights that are a function of American federal citizenship, such as the right to be heard in courts of justice and the right to diplomatic protection." The Supreme Court has specifically rejected the claim that the clause protects rights that result from state citizenship.

Unfortunately, a considerable amount of federal judicial oversight of the states has been created through interpretations of the clause over the years. However, many 21st century Progressives are now seeking additional federal oversight of state and local laws. "Their strategy in this regard is to refashion the privileges or immunities clause as a new and essentially unlimited bill of rights within the 14th Amendment. The practical consequences of this would be to authorize federal judges to impose an ever broader and more stultifying uniformity upon the nation."

You can clearly see how a major departure from the Constitution is being contemplated. Furthermore, the plan is to accomplish it without any formal Constitutional amendment process and without the knowledge of the American people.

- <u>Negative Liberties versus Positive Rights</u>

"For the 21st century constitutionalist, perhaps the greatest virtue of redefining the privileges or immunities clause is the prospect of transforming the Constitution from a guarantor of "negative liberties" into a charter of "affirmative government." The Constitution defines what the government cannot do to you – e.g., unreasonable search and seizure – it doesn't guarantee rights to such things as education, health care, an appropriate job, and affordable housing.

President Obama clearly supports this transformative change. He criticized the Warren Court of the 1950s and 1960s by saying in a radio interview, "The Warren Court…didn't break free from the essential constraints that were placed by the Founding Fathers in the Constitution…that generally the Constitution is a charter of negative liberties, says what the states can't do to you, says what the federal government can't do to you, but it doesn't say what the federal government or the state government must do on your behalf."

The idea of negative liberties is, of course, the foundation on which our form of government is built. The Founders designed into the Constitution limits on the power of the federal government. This approach is consistent with the key responsibility of government which is to maximize liberty, as discussed in the previous chapter.

There is nothing in the Constitution about the kinds of positive rights listed above. Positive rights are the brainchild of Progressives and activists and provide for them the justification for more involvement of the federal government in our society.

Unfortunately, the courts have supported these distortions in the meaning of the Constitution and have created some of their own – see the earlier discussion of the Supreme Court's decision in the case of Roe versus Wade. The courts are complicit with the Congress in allowing the Constitution to be amended without going through the formal amendment process.

This is, of course, just the opposite of the role for which the judiciary was created by the Founders. The judiciary is supposed to determine the constitutionality of the measures passed by Congress, reject those which are clearly unconstitutional based on the Constitution and its amendments, and require that the amendment process be used to make constitutional those ideas which are currently unconstitutional.

I believe that the creation and judicial support for positive rights is the greatest threat to liberty in the history of the Republic. The attempt to transform the design of the Constitution to make it proactive with respect to "positive rights" without formal amendment cannot be allowed to continue.

There is nothing in the Constitution which specifies positive rights. If this potential transformation of the Constitution is not communicated to the American people and summarily rejected, equality of condition will have made the biggest advance in the history of the nation.

- <u>State Action</u>

"A barrier posed by both the due process and privileges or immunities clauses, and viewed as anachronistic by 21st century constitutionalists, is the requirement of state action as a precondition of the enforcement of rights. In 1883…the Supreme Court asserted that these provisions of the 14th Amendment prohibited only the abridgment of individual rights by the *state*. In other words, "It is state action of a particular character which is prohibited…".

"However, for advocates of 21st century constitutionalism, if fairness and equity are to be achieved, the Constitution must become more like a general legal code – applicable to both public and private institutions."

Consider the case of Hillsdale College, "…the embodiment of a thoroughly private institution … government officials have sought to justify the imposition of federal rules and regulations upon Hillsdale by characterizing the college as the equivalent of a state entity on the grounds that it received public grants-in-aid. We have witnessed a steadily more aggressive effort by governmental regulators to treat private institutions as the equivalent of the state, and thereby to extend public oversight".

Despite the fact that Hillsdale does not accept public funding of any kind, this interpretation would mean that "… Hillsdale's policies concerning such things as tuition, admissions, faculty hiring, curriculum, and discipline will each have to pass the scrutiny, and receive the imprimatur, of judges".

So here we have the bizarre case of an attempt by the government: (1) to claim that an educational institution is a state (which doesn't make any sense at all), and (2) then to claim that it can regulate it because of its presumed power to regulate states (which is basically unconstitutional). This is another attempt to extend public oversight in ways which are not to be found in the Constitution.

- Political Questions

John Marshall in the case of *Marbury v. Madison* in 1803 said that "Questions in their nature political... can never be made in this Court." However, "In areas that were once viewed as inappropriate for judicial involvement, federal courts have begun to assert themselves in an unprecedented and aggressive manner."

For example,"... in the last few years, the Supreme Court in a series of 5-4 decisions has overruled decisions by both the legislative and executive branches regarding the treatment of enemy combatants. "These decisions have been made despite the facts that these individuals have been captured in combat, without any indication that they represented any country and held as prisoners of war in a war which was duly authorized and conducted under the provisions of the Constitution." The Court has ruled that they possess the constitutional right to challenge their detentions in federal court.

These kinds of modifications to the Constitution are being implemented at all levels of the judiciary on all kinds of issues. "As a result, every policy debate taking place within government, at every level will become little more than a prelude for judicial resolution".

- Ninth Amendment

The Ninth Amendment says, "The enumeration in the Constitution of certain rights, shall not be construed to deny or disparage others retained by the people." The conventional understanding is that, "it was written to dispel any implication that by the specification of particular rights in the Bill of Rights, the people had implicitly relinquished to the new federal government rights not specified."

However, "many constitutionalists understand this amendment to say that there is some unknown array of unenumerated rights that lie fallow in the Constitution, waiting only to be unearthed by far-sighted judges." A professor at the Stanford Law School has suggested that the Ninth Amendment constitutes a 'license to constitutional decision makers to look beyond the substantive commands of the Constitutional text to protect fundamental rights not expressed therein." [such as abortion].

"The principal purpose [of the Ninth Amendment] was to prevent an extension of federal power, not to provide an open-ended grant of judicial authority that would have the opposite effect."

- Transnationalism

A huge controversy has developed between "transnationalists" and "nationalists".

"Transnationalists believe that international and domestic law are merging into a hybrid body of transnational law, while so-called nationalists persist in preserving a division between domestic and foreign law that respects the sovereignty of the United States law. Transnationalists believe that domestic courts have a critical role to play in incorporating international law into domestic law, while so-called nationalists claim that only the political branches are authorized to domesticate international legal norms."

Not only would American judges have to be asked to rely on foreign law to give meaning to their interpretations of the Constitution, but also the US would be bound by international treaties and agreements which have never been ratified and enacted into US law. International conduct of the US would be judged by international tribunals with no history of the development of US law. They would be charged with balancing the interests of a broad range of countries with legal systems which have little resemblance to our own.

However, it should be clear that this country has a Constitution on which our system of laws is based, and US judges are necessarily and appropriately limited to interpreting what our Constitution says, as opposed to what world opinion says. Anything else would be nonsensical and incompatible with the logic of the Founders in creating the unique document which has been so pivotal in the development of one of the greatest civilizations the world has ever seen.

Conclusion

With the discussion in this Chapter, we now have added the third leg of the stool supporting political life in America:

- If people join a social contract society, they do it because they are seeking liberty (Chapter 5).
- Within the society, people have the maximum liberty if the role of government is limited (Chapter 5).
- The rule of law protects that liberty from encroaching government, on the one hand, and other people, on the other (this Chapter).

This is not utopia on earth, but it is the most ideal society the world has ever seen.

The problem is that two of the legs of the stool are under attack. As I have tried to illustrate, the role of government is constantly expanding, which, by definition, means that liberty is shrinking. Furthermore, the rule of law is being eroded by Progressives and others who believe that the Constitution is a "living" document subject to an ever-changing interpretation and application.

If these attacks are allowed to continue, the results are predictable. The American civilization will collapse from the inside, even if external threats are addressed. However, it is proving difficult enough to address external threats, both military and economic, and the concept of a "living constitution" puts an additional burden on the structure of our society.

The answer to the question of what to do at this point is self-evident by looking at the legs of the stool. We have to:

- Promote individual liberty, both as a concept and in fact
- Limit the role of government, so that liberty can flourish
- Apply a rule of law which is well-grounded in the Constitution and which does not rely on any other source for its authority.

We cannot always anticipate the host of external military and economic threats with which we are and will be confronted. However, we can be in the best possible position to deal with them and preserve out remarkable civilization if we adopt the principles and the related necessary actions which are part of the Second American Revolution.

Bottom Line

- People sign up for a social contract to pursue liberty, create a government which has as one of its primary purposes maximizing liberty, and create a rule of law to protect them from the government and each other.
- The Rule of Law in this country starts with the Constitution and the Bill of Rights.
- There is a formal amendment process outlined in the Constitution itself, and attempts to alter it through legislative and judicial fiat are, by definition, illegitimate.
- The concept of a "living" Constitution is therefore illegitimate, because it relies on everything but the formal amendment process to achieve its objectives.
- There are at least 6 major issues which will be part of the upcoming Constitutional debate.
- One of the most important of these is the effort to transform the Constitution from a guarantor of "negative liberties" (e.g., unreasonable search and

seizure) to a provider of "positive rights" (e.g., health care, a good job, and affordable housing) without going through the formal amendment process.
- Two of the three legs of the three-legged stool described in the first bullet point above (liberty and the rule of law) are under vigorous attack.
- To Restore the Future, we must re-establish all three legs of the stool on firm Constitutional ground.

CHAPTER 7
NATIONAL DEFENSE

> WE THE PEOPLE of the United States, in Order to form a more perfect Union, establish Justice, insure domestic Tranquility, provide for the common defence, promote the general Welfare and secure the Blessings of Liberty to ourselves and our Posterity do ordain and establish this Constitution for the United States if America.
>
> United States Constitution

One of the primary responsibilities of government, as indicated in the Constitution, has to be national defense. Without a strong national defense, concepts like Tranquility, Welfare, and Liberty are meaningless, because they don't exist.

Background

James Madison in Federalist #41 observed, "Security against foreign danger is one of the primitive objects of civil society. It is an avowed and essential object of the American Union".

"The pursuit of happiness is a natural right of liberty, but safety is the initial requirement of the pursuit. Among the many objects to which a wise and free people find it necessary to direct their attention, that of providing for their safety seems to be first" John Jay wrote in Federalist #2.

Alexander Hamilton said in <u>Federalist #8</u>, "Safety from external danger is the most powerful director of national conduct. Even the ardent love of liberty will, after a time, give way to its dictates." What Hamilton is saying is that <u>in extremis</u> people will give up a lot to survive, if necessary.

Hamilton writes in <u>Federalist #34</u>, "to judge from the history of mankind, we shall be compelled to conclude that the fiery and destructive passions of war reign in the human breast with much more powerful sway than the mild and beneficent sentiments of peace; and that to model our political systems upon speculations of lasting tranquility, is to calculate on the weaker springs of the human character."

George Washington felt very strongly about the importance of national defense at the federal government level as a result of his experience with conducting the Revolutionary War under the Articles of Confederation. Washington was given full responsibility to fight the War, but the persistent lack of supplies left his army in such a depleted condition that we almost lost the War at several points. The states, under the Articles, were not obligated to furnish supplies, or even personnel, and they reluctantly provided wholly inadequate amounts of both.

National defense is not just protecting our soil from attack. US interests exist in all parts of the world, and, by extension, they need to be protected as well. Furthermore, it is important to keep in mind that, "The right of a sovereign nation to preserve itself is not merely passive or defensive. Sovereignty also entails a protective right to eliminate threats".[1] Thus, we are justified in taking pre-emptive action against enemies if we believe that they are planning to attack us.

The importance of national defense to the Founders is self-evident, and it was a major focus of their thinking. The problem is that the Founders were also committed to limited government.

In fact, the issue of a "standing army" was one of the important issues separating the Federalists (Washington and Hamilton) from the Anti-Federalists (Jefferson and, later, Madison). There was no national army at the time of the War, and personnel were provided to Washington at the discretion of the states from their militias.

"Given their view that the federal government they were bringing into existence constituted the biggest threat to their freedom and well-being, constantly on the minds of our ancestors was the primary means by which governments had historically subjected their people to tyranny – through the use of the government's military forces."[2]

This view is understandable. "Wrenching memories of the Old World lingered in the 13 original English colonies along the eastern seaboard of North America, giving rise to deep opposition to the maintenance of a standing army in time of peace. All too often the standing armies of Europe were regarded as, at best, a rationale for imposing high taxes, and, at worst, a means to control the civilian population and extort its wealth."[3]

Nevertheless, the Founders recognized the failings of the Articles of Confederation and the world situation in which they were vulnerable to attack by foreign powers, and they were forced to accept the idea of a standing army. The Constitution was the answer to controlling a standing army, by limiting the powers of the federal government to just those enumerated in the document.

Democracies find it extremely difficult to prepare for the next war, because in times of peace, priorities change. Spending on defense can take a back seat relative to other priorities. This was the situation with Pearl Harbor, for example. At the time, we were poorly prepared to wage war in both Europe and the Pacific, but that is not unusual. Our genius was, and is, that we can build up dramatically quickly a war machine which is the envy of the world and historically sufficient to win any war.

We have to have a national defense capability which is designed to respond quickly to threats anywhere in the world at some minimum level, on the one hand, and which at the same time stands ready to ramp up that capability in the face of greater threats than had been expected, on the other.

This is a difficult tradeoff to make in a democracy, and people have debated what it should be ever since the Constitution was signed, with no clear resolution.

Current Situation

As can be seen in Figure 7-1, which was also shown in Chapter 2, Obama's budget would return defense spending as a percentage of GDP by 2019 to levels comparable to those which existed before 9/11. Furthermore, this level is not only well below the average for the last 45 years, it would match the low for this entire period.

Figure 7-1

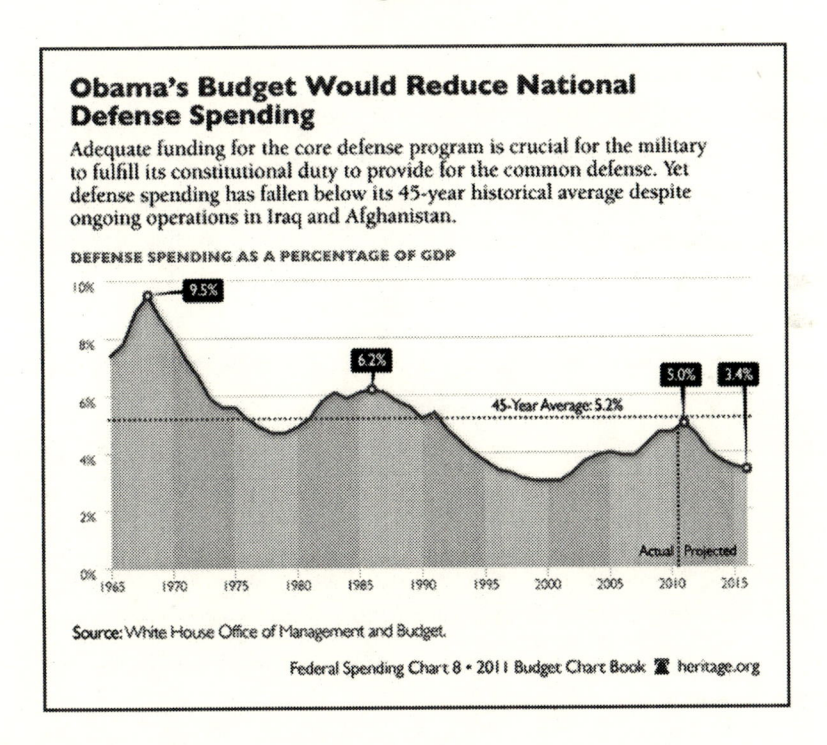

Obama's Budget Would Reduce National Defense Spending

Adequate funding for the core defense program is crucial for the military to fulfill its constitutional duty to provide for the common defense. Yet defense spending has fallen below its 45-year historical average despite ongoing operations in Iraq and Afghanistan.

DEFENSE SPENDING AS A PERCENTAGE OF GDP

1965: 9.5%
1985: 6.2%
45-Year Average: 5.2%
2010: 5.0%
2015: 3.4%

Source: White House Office of Management and Budget.

Federal Spending Chart 8 • 2011 Budget Chart Book • heritage.org

The average obviously includes periods of high spending, associated with major wars, such as the Vietnam War in the 1960s, as well as periods when politicians were trying to achieve a "peace dividend", such as the 1990s, by deemphasizing defense spending.

It appears from Figure 1 that the Obama administration is once again trying to achieve a "peace dividend". I don't think that pursuing this strategy makes sense in the first place, because we should maintain a constant level of military readiness, even if there are no immediate threats on the horizon. However, if there is any justification at all, it has to be because we are at peace and can afford to spend money on other priorities. Are we at peace at this point?

The defense strategy of this country at the moment is to follow the maxim that the best defense is a good offense. It makes sense to take the war on terror to the terrorists to keep them occupied and ultimately defeat them, rather than waiting for them to attack us and try to deal with them at that point. This is why we were in Iraq and are in Afghanistan today with large numbers of conventional forces.

"The range of potential missions facing today's military is vast. While winning the wars in Iraq and Afghanistan remains a central mission, regional combat commanders must also respond to humanitarian disasters, protect sea lines of communication and free trade, deter rogue states through a credible extended deterrence posture, and hedge against the future uncertainty that accompanies the rise of powers like China and Russia."[4]

"Two primary components determine a strong military: the quality of its service members, and the modern, technologically advanced equipment available to them."[5] American military personnel in terms of training and capability are the best in the world. However, equipment is another story.

"The collective decisions by Congress and both Democratic and Republican Presidents over the last 15 years have left the U.S.military using equipment that is extremely old and, in many cases, outdated."[6] Consider the following statistics about the age of major weapons systems:

- "Air Force tactical aircraft – over 20 years old
- Navy and marine Corps tactical aircraft – over 15 years old
- Bradley fighting vehicles [tanks] – 17 years old
- UH-1 Huey helicopters – 35 years old

- B-1 Lancer bomber – over 20 years old
- Army M-113 armored personnel carriers – 27 years old
- Ticonderoga-class cruisers – nearly 20 years old"[7]

This equipment is so out-of-date that it can't possibly be considered as competitive on the world stage. Furthermore, it is so old that it is probably difficult to get spare parts to keep the equipment operating in the first place. There is more.

Jim Talent, a distinguished fellow at the Heritage Foundation and former U.S. senator explains: "The Navy has fewer ships than at any time since 1916. The Air Force inventory is smaller and older than at any time since the service came into being in 1947. The Army has missed several generations of modernization..."[8] "Not only have we lost our enthusiasm for the exploration of space, we have retreated on the seas. Up to 30 ships, the largest ever constructed, will soon come off the ways in South Korea. Not only will we neither build, own, nor man them, they won't even call at our ports, which are not large enough to receive them."[9]

Furthermore, "We have the smallest navy in almost a century, declining in the past 50 years to 286 from 1,000 principal combatants... Even if the capacity of the whole navy could be packed into a small number of super ships, they could be in only a limited number of places at a time, and the loss of just a few of them would be catastrophic."[10]

"The overall effect of recent erosions [of naval capacity] is illustrated by the fact that 60 ships were commonly underway in America's seaward approaches in 1998, but today, despite opportunities for the infiltration of terrorists, the potential of weapons of mass destruction, and the ability of rogue nations to sea-launch intermediate and short-range ballistic missiles – there are only 20"[11]

This is a pathetic and very dangerous situation. It is the responsibility of the government to maintain an adequate response capability to conventional warfare, and these figures demonstrate that, independent of how that level of military capability is defined, our equipment is so out-of-date that it is hard to believe that we can achieve that objective.

Even if we were better prepared for conventional warfare, the primary political/military threat we face may be the development by a rogue nation of a nuclear capability with the potential to create nuclear missiles.

"North Korea is currently developing a long-range ballistic missile that could soon carry a nuclear warhead all the way to Alaska or California. Iran already has missiles that can reach Europe, and could soon acquire nuclear weapons. These countries could share their missile and nuclear technologies with terrorists, who would in turn be able to directly threaten New York City and other American cities with short-range missiles"[12]

"The truth is brutal – no matter where on earth a missile is launched from it would take 33 minutes or less to hit the U.S. target it was programmed to destroy."[13] We have to create the technology to intercept any missile which is aimed at the United States if we want to make it a priority. However, unfortunately, the Obama administration is cutting back in spending on missile defense, just as is the case in other areas.

The preceding discussion was about long-range ballistic missiles. These are the ones which can reach American cities in 33 minutes. However, the same cutbacks are taking place in tactical missile defense, which is concerned with short-to-medium range ballistic missiles.

"The U.S. Department of Defense has decided to stop funding for the Medium Extended Air Defense System (MEADS), a ground-based terminal ballistic missile defense (BMD) developed jointly by the United States, Italy, and Germany... The MEADS program is designed to protect the United States' homeland, allies, and forward-deployed troops against a wide range of threats, including the next generation of tactical ballistic missiles."[14]

This is madness. Surely, the responsibility of the government to provide for the "common defence" extends to protecting us from something which could destroy an important part of our civilization in a matter of minutes?

Ronald Reagan said, "It's up to us in our time to choose and choose wisely between the hard but necessary task of preserving peace and freedom, and the temptation to ignore our duty and blindly hope for the best while the enemies of freedom grow stronger by the day"[15]

How did we get into this situation? As suggested earlier and as demonstrated in Figure 7-2[16], defense spending has been squeezed out over the last 40 years by the massive growth of entitlements.

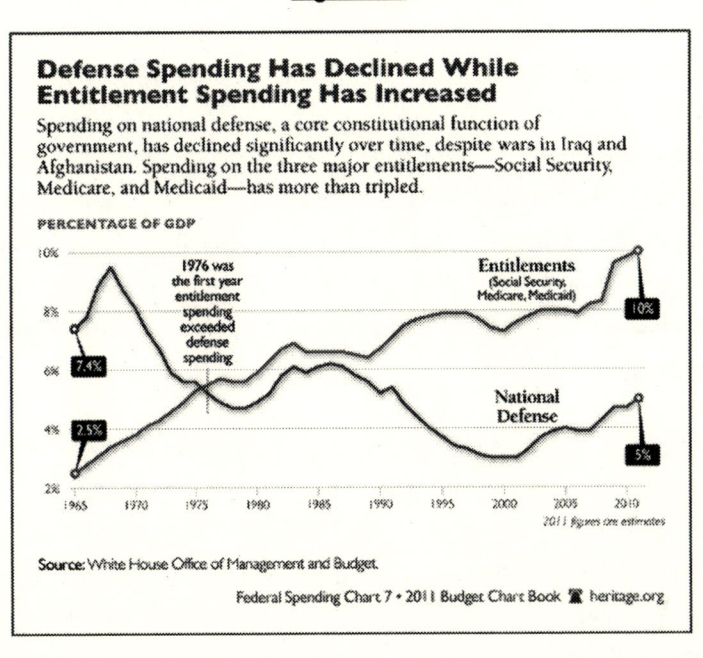

Figure 7-2

Defense Spending Has Declined While Entitlement Spending Has Increased

Spending on national defense, a core constitutional function of government, has declined significantly over time, despite wars in Iraq and Afghanistan. Spending on the three major entitlements—Social Security, Medicare, and Medicaid—has more than tripled.

PERCENTAGE OF GDP

1976 was the first year entitlement spending exceeded defense spending

Entitlements (Social Security, Medicare, Medicaid): 10%

National Defense: 5%

7.4% → 2.5% (1965)

2011 figures are estimates

Source: White House Office of Management and Budget.

Federal Spending Chart 7 • 2011 Budget Chart Book ■ heritage.org

Defense spending has declined from 7.4% of GDP in 1965 to 5.0% in 2010, and this is after the buildup in recent years of spending for Iraq and Afghanistan. This is a decline of about one-third. In the same time period, spending on entitlements has increased from 2.5% in 1965 to 10.0% in 2010, an increase of 4 times.

As noted in Chapter 4, Social Security is designed to be self-funding, so that it does not compete directly for spending with national defense. However, it is still part of the total government influence on economic activity.

• 147 •

Most of the entitlements are not discretionary after they have been implemented, which makes the situation even worse. How would you, for example, double the percentage of GDP being spent on national defense if the proportion of GDP spent on non-discretionary spending for things like entitlements is constantly rising and extremely difficult to roll back?

This is one of the greatest possible indictments of the Federal government today. It is impossible to make the argument that the world is a less dangerous place and that pursuing a peace dividend is in any way justified. Entitlements are obviously totally unrelated to defense spending and in no way contribute to even a minimal level of military readiness. Yet they have been allowed to grow in a way that has significantly reduced the available resources for national defense spending.

If we believe with the Founders that one of the primary responsibilities of government is to "provide for the common defence", this situation makes no sense at all. Does it make sense to compromise national security in the interest of paying benefits to people who are "entitled" to them in a process which has no constitutional justification? Without an adequate defense capability, entitlements are totally irrelevant anyway.

The first priority should be to determine what minimum level of spending on national defense is required. We need to maintain an adequate level of readiness and then decide how to spend what is left over, rather than the other way around. It is hard to argue that the current level of spending on national defense is determined in this way. It looks as if spending on defense is an unfortunate and dangerous residual.

This is bad enough, but the situation is actually even worse than portrayed here. Conventional warfare against conventional attack is not the only thing for which we need to be prepared. In the modern era, there are many non-conventional threats, for which we must also be prepared.

Terrorism is defined both by the type of terrorism and by the attack method which is utilized. For example, "state-sponsored" terrorism in one of the most important types. As far as the U.S. is concerned, "State Sponsors of Terrorism" is a designation applied by the United States Department of State 'to nations which are designated by the Secretary of State 'to have repeatedly provided support for acts of international terrorism"[17]

There are currently four countries on the list: Cuba, Iran, Sudan, and Syria. Iraq, Libya, North Korea, and South Yemen were on the list, but they have been removed. For countries on the list, there are sanctions imposed by the U.S., such as a ban on arms-related sales, prohibitions on economic assistance, and imposition of miscellaneous financial and other restrictions.

This, of course, is the big picture with which many people are familiar, and keeping track of what is going on in these countries and the countries they influence is a very big job. However, there is also the problem of keeping track of methods which could be used by terrorists to attack. These provide another dimension of the terrorist threat, because they cut across all the identified state-sponsored terrorist groups, as well as many others.

All terrorism experts do not agree on the definitions of the types of terrorism. However, what I think they all do agree on are the types of terrorist attack methods. In fact, "under the National Infrastructure Protection Plan, the Department of Homeland Security has the responsibility to…analyze information about terrorist attack capabilities, goals, and objectives to assess the potential terrorist attack methods which might be used against…the nation's critical infrastructure."[18]

Their analysis is based on "a defined set of 15 attack methods that were identified based on known terrorist capabilities, analysis of terrorist tactics, techniques, and procedures, and intelligence reporting on assessed, implied, or stated intent to conduct an attack"[19].

If you consider the number of specific kinds of threats which you get by combining just four state-sponsored terrorist countries with 15 implementation options, you get 60 different combinations of country/delivery mechanism (Iran/nuclear, Syria/biological, Cuba/cyber, etc.). This is a very large number.

By the way, you will note that Homeland Security, the Center for Disease Control, and other groups each have some responsibility for defending against these threats. It doesn't matter which government agencies are involved, the issue is still one of national defense.

Some of the 15 identified attack methods are the following, and I would like to discuss each of them briefly:

- Biological
- Cyber
- Nuclear

"Bioterrorism refers to the intentional release of toxic biological agents to harm and terrorize civilians... The U.S. Center for Disease Control has classified the viruses, bacteria, and toxins that could be used in an attack. For example, category A includes:

- Anthrax
- The plague
- Botulism
- Smallpox[20]

"The act of bioterrorism can range from a simple hoax to the actual use of these biological weapons, also referred to as agents. A number of nations have or are seeking to acquire biological warfare agents, and there are concerns that terrorist groups or individuals may acquire the technologies and expertise to use these destructive agents. Biological agents may be used for an isolated assassination, as well as to cause incapacitation or death to thousands. If the environment is contaminated, a long-term threat to the population could be created."[21]

Biological warfare has been used for millennia, from the ancient Greek archers who used contaminated waste on their arrows, to gifts to the American Indians by the British in the French and Indian War of blankets which had been used by smallpox victims in an effort to spread the disease, to the Viet Cong who used punji sticks dipped in feces to cause significant infections in their victims after they were injured.

However, the use of biological agents primarily against civilians in the modern era is something largely unprecedented. The fear is heightened by the realization that some of these biological agents are relatively easy to manufacture or purchase, and there are a myriad of ways in which to deliver them.

What is cyber terrorism and why is it important? "The online threats facing America read like an ever-expanding encyclopedia of dangers to the freedoms, prosperity, and security of all Americans. Cybersecurity has become a crucial component of national security... According to the Defense Science Board, not only do cyber attacks represent a general threat, but military and DOD operations are, to a significant extent, susceptible to their effects."[22]

As part of its effort to organize its activities, "the Department of Homeland Security has identified 18 sectors of the economy as the nation's critical infrastructure and key resources. As one would expect of a comprehensive list, it covers everything from transportation to the defense industrial base. It also includes energy, financial systems, water, agriculture, and telecommunications. The remarkable thing is that virtually all of the sectors now substantially depend on cyber systems."[23]

"A cyber attack comprises the actions taken through computer networks to disrupt, deny, degrade, destroy, or manipulate information in computers or computer networks, or the computers or networks themselves."[24]

"No good data exist on precisely how many cyber intrusions occur annually. The number is so great that in 2004, the U.S. government stopped reporting the number of known intrusions, which in 2003 exceeded 100,000. Most experts presume that the number today is an order of magnitude larger."[25]

Cyber terrorism is something most people have probably not thought much about. However, if they pause to think about it, they will instantly realize the nature of the problem. "Severe cyber attacks could disrupt, deny, destroy, or allow hackers to exploit systems and networks essential to the functioning of critical U.S. infrastructure with potentially devastating effects on economic security, the environment, national security, and public health and safety."[26]

"Cyber attacks occur in directed and non-directed forms. Directed attacks target specific computer systems or networks or use those systems or networks to attack other targets. Non-directed attacks are not targeted at a particular entity or sector, but can cause widespread disruptions to systems and networks."[27]

Cyber attacks are not just theoretical. "The Pentagon on [July 14, 2011] revealed that in the spring it suffered one of its largest losses ever of sensitive data in a cyber attack by a foreign government [thought to be China]. It's a dramatic example of why the military is pursuing a new strategy emphasizing deeper defenses of its computer networks, collaboration with private industry and new steps to stop 'malicious insiders.'"[28]

"Among the disturbing emergent concerns is that enemies could create a catastrophic failure in Supervisory Control and Data Acquisition (SCADA) systems, which monitor and control most U.S. infrastructure. Such an attack could cause power outages, spark explosions, and unleash fuel spills."[29]

Think about how important the electric grid is to the country or how often you use the internet. What would happen if we had an extended power outage and the internet were not available for a long time? An increasing number of people use electronic banking, and what would happen if they couldn't access their bank? Many people hold stocks, and what would happen if they were suddenly completely illiquid because the stock exchanges couldn't function? How would you feel if the electronic records for your retirement plan were unavailable or lost?

We rely on alarms and lights to keep us safe, and we assume that everything will always work the way it is supposed to work. How safe would we be if they did not work? Telephones, transportation networks, and defense systems would be disabled. We are prisoners of our technological achievements, and the situation gets potentially worse with each advancement.

There is no question that we are extremely vulnerable to cyber terrorism, yet how many people are aware of this vulnerability or what is being done to reduce it? Just as is the case with all of the 15 threats which were referenced above, cyber terrorism should be a high priority for those we elect to protect us. If it is not part of the common discussion, how can we know that it is?

It is important to note that the recognition of the critical importance of cyber terrorism has led the Pentagon to announce that cyber attacks can now be declared "an act of war". Cyber terrorism obviously cannot be any more important than this.

The third type of terrorism I selected from the 15 identified above is nuclear terrorism. What does this mean? "Nuclear terrorism' refers to a number of different ways nuclear materials might be exploited as a terrorist tactic. These include attacking nuclear facilities, purchasing nuclear weapons, or building nuclear weapons or otherwise finding ways to disperse radioactive materials"[30]

A terrorist attack on a nuclear facility would be a disaster. "In addition to the reactors themselves, nuclear power plants harbor enormous quantities of radioactive materials in spent fuel pools. On average, these spent fuel pools contain five times as much radioactive material as the reactor core, and they are housed in corrugated steel buildings even more vulnerable to attack than the reactor containment buildings. The vulnerability of nuclear power plants is highlighted by reports that 47% of US nuclear power plants failed to repel mock terrorist attacks conducted by the Nuclear Regulatory Commission during the 1990s."[31]

You may not have thought about it, but a disaster like the recent nuclear meltdown in Japan, which was caused by a tsunami, could have been the result of a terrorist attack as well. "The nuclear industry in many countries is much less prepared to cope with security incidents than with accidents...the chance that the next big radioactive release will happen because someone wanted to make it happen may well be bigger than the chance that it will happen purely by accident."[32]

It is clear that defense against bioterrorism, cyber terrorism, nuclear terrorism, and the other terrorist threats should be part of our national defense establishment.

These threats are monitored by a number of different government agencies, but that is a triumph of form over substance, unless there is some overall coordination. You recall that this was lacking in the case of homeland security on 9/11. Defense is defense, and all of the threats related to types of terrorist attack have to be considered to be part of the overall responsibility of government.

How much of its national output a country decides to spend is one indication of how seriously it takes the importance of national defense. Some critics argue that the historical average for the US of 5.2% is too high and far more than other countries spend. This, of course, is an attempt to develop a rationale for cutting defense spending.

The Stockholm International Peace Research Institute has developed an extensive data base of how much all the countries in the world spend on defense as a percentage of GDP. The data for some countries I have selected are for 2009, which is the latest year for which data are available, are shown below:

Figure 7-3

Israel	6.3%
Iraq	5.4
US	4.7
Russia	4.3
UK	2.7
China	2.2
Germany	1.4
Japan	1.0

Given the size of our country, the vast interests we have around the world, the fact that we participate in collective defense organizations for which we have to provide the lion's share of the funding, the enormous increase in the threats with which we are confronted, and state of our military readiness from the standpoint of equipment, it is difficult to make an argument that the long-term average of 5.2% of GDP which we have historically spent is too high. If anything, the argument should be that it is too low.

Conclusion

At the beginning of this chapter, I documented the dismal news of the decline in defense spending in the Obama budget from 5.2% of GDP to 3.4% of GDP over the next ten years. This chapter has presented evidence that this reduction is ill-conceived and dangerous. This government is not living up to its constitutional mandate to provide for the "common defence", and every day which goes by without recognizing this problem is another day in which we are rolling the dice with our survival as a nation.

This is as much an indictment of the Obama administration as were the grievances against King George listed by Thomas Jefferson in the Declaration of Independence, and the outcome should be the same.

Defense Secretary Robert Gates left the Department of Defense after 4 years and two administrations, and he has issued a farewell warning: "Mr. Gates has warned against cuts to weapon programs and troop levels that would make America vulnerable in "a complex and unpredictable security environment".[33] Gates also said. "But make no mistake, the ultimate guarantee against the success of aggressors, dictators, and terrorists in the 21st century, as in the 20th, is hard power – the size, strength, and global reach of the United States military."[34]

Furthermore, "Gates acknowledged it only in passing this week, but the reality is that the entitlement state is crowding out national defense."[35] This, of course, is the same argument which I was making earlier in this chapter.

Finally, "Last year, Mr. Gates said that the Pentagon needs 2-3% budget growth in real terms (after inflation) merely to sustain what it's doing now, but it could make do with 1%. The White House gave him 0%."[36]

Of course, no one can say what the exact level of defense spending should be. However, this chapter has demonstrated that there is no reason to think that the long-term average is not a good place to start.

It was George Washington who said, "There is a rank due to the United States among nations which will be withheld, if not absolutely lost, by the reputation of weakness. If we desire to avoid insult, we must be able to repel it; if we desire to secure peace, one of the most powerful instruments of our rising prosperity, it must be known that we are at all times ready for war."[37]

This statement is as true today as it was when George Washington said it. While war had a certain meaning to him, the nature of war is far more complex in today's world than it was in his. However, that doesn't change anything. His statement applies to war on all fronts.

If you were asked the following two questions, what would you say: (1) do you feel safer than you felt 10 years ago?, and (2) if present conditions continue, do you expect to feel safer 10 years from now? If your answer is "no" to either of both questions, as I suspect it is for many people, you must both support and participate in the Second American Revolution, which is the only way to Restore the Future.

Bottom Line

- National defense is one of a very limited number of Federal government responsibilities specified in the Constitution.
- Without a strong national defense capability, nothing else matters.
- Republics find it difficult to maintain an adequate defense capability in peacetime.
- The Obama administration is planning to cut defense spending as a percent of GDP to an historic low.
- The administration is specifically cutting investment in missile defense.
- Our military equipment is dangerously out-of-date, and force levels, particularly for the navy, have declined significantly.
- Defense spending has been squeezed over the last 40 years by the astronomic growth in entitlements.
- Our national defense capability must be prepared to deal with at least 4 countries classified as "State Sponsors" of terrorism and as many as 15

different types of terrorism, including bioterrorism, cyber terrorism, and nuclear terrorism.

- Our level of defense spending must be determined before spending on anything else by a careful analysis of the threats with which we are confronted and the resources necessary to overcome them, rather than being a residual after spending on other areas, including entitlements, is determined.

CHAPTER 8
EDUCATION

A mind is a terrible thing to waste

United Negro College Fund

The world's greatest concentration of PhDs is in Seoul, Korea, and half of Americans can't even find Seoul on a map[1]

Pascal Forgione
Former U.S. Commissioner of Education Statistics

Education is a critically important issue for the country at any time, and this is particularly true at this point in our history, as I will show below. However, experts don't agree on what the quality of education today is in the United States or on what to do about it if it needs to be improved.

This country must have very clear objectives about the education we want to have and what the best way to deliver it is. As far as I can tell, such objectives do not exist. Even if they did, we do not have an agreed-on set of tools to evaluate the extent to which the objectives are being achieved. Having objectives without being able to know whether they are being achieved is a triumph of form over substance.

I am obviously not an educator in the conventional sense. However, setting objectives for public education is not rocket science and does not require an advanced degree in education.

In my opinion, appropriate objectives are to produce: (1) "educated" men and women in the classical sense with a broad range of knowledge (mathematics, science, reading, writing, grammar and punctuation, history, civics, geography, foreign languages, and problem solving), which will make them best prepared to live in any situation they encounter in an increasingly complex world, and (2) extraordinarily high scores on standardized tests compared to their peers from other parts of the world, which demonstrate that our students are competitive on the world stage.

Setting objectives is no more complicated than that. These objectives are easy to understand, easy to evaluate in a non-subjective way, inspirational to both parents and children, and galvanizing for the people of our country who are unclear about what our educational objectives are or should be.

This chapter is designed to provide some background information on who in the country is responsible for education, a detailed analysis of the quality of our education today, and some key recommendations to dramatically improve the tangible results of our educational system.

Background

A lot of people don't know who is responsible for education in this country. Is it the federal government, the states, local governments, or some combination? Making this clear is critical for understanding the rest of this chapter.

"The right to a free public education is found in the various state constitutions and not in the federal constitution. Every state has a provision in its constitution, commonly called the 'education article', that guarantees some form of free public education, usually through the twelfth grade. [It is important to note that] in San Antonio Independent School District v. Rodriguez, the Supreme Court in 1973 held that education is not a 'fundamental right' under the U.S. Constitution.

Thus, as a matter of constitutional law, the founding fathers left it to the states to decide whether to provide an education or not, and, if deciding to provide one, determine at what level of quality."[2]

This makes it clear that education is a fundamental responsibility of each state government. Remember the discussion in Chapter 6 about negative liberties, which are specified in the Constitution, and positive rights, which are not. The federal government has no role, and education is not a positive right. However, this is not how things have worked out.

"Not only does the federal constitution confer no right to education, it does not even explicitly empower the U.S. Congress to legislate on the subject. Most federal legislation is therefore enacted under the 'spending clause' of the Constitution, which gives Congress the authority to tax and spend for the general welfare. Since federal grants to the states may be conditioned upon the state's adoption of certain legal and regulatory structures, the federal government has been able to exercise substantial authority over K-12 education policy."[3]

How can this happen, given the constitutional provisions? When the states establish an education policy, it must be consistent with constitutional rights, such as equal protection under the law, and the First Amendment's requirement for the free exercise of religion. Thus, even though it does not specifically give the federal government a role, the Constitution still has a major effect on American education. The extent of the federal government's role will be examined later in this chapter.

Current Situation

I would like to start by describing the performance of U.S. students on standardized tests compared to the performance of students in other countries. There are at least two major organizations which conduct these kinds of tests on a scheduled basis.

"The Programme for International Student Assessment (PISA) is a worldwide evaluation of 15-year old school pupils' scholastic performance, performed first in 2000 and repeated every three years. It is coordinated by the Organization for Economic Co-operation and Development (OECD). PISA aims at testing literacy in three competence fields: reading, mathematics, and science.

The PISA mathematics literacy test asks students to apply their mathematical knowledge to solve problems set in various real-world contexts. To solve the problems, students must activate a number of mathematical competencies as well as a broad range of mathematical content knowledge.

In the reading test, 'OECD/PISA does not measure the extent to which 15-year-old students are fluent readers or how competent they are at word recognition tasks or spelling'. Instead, they should be able to 'construct, extend, and reflect on the meaning of what they have read across a wide range of continuous and non-continuous tests.

One of the other major tests is the Trends in International Mathematics and Science Study (TIMSS), which focuses on math and science but not reading. TIMSS measure more traditional classroom content such as an understanding of fractions and decimals and the relationship between them (sometimes called "curriculum attainment"). PISA claims to measure education's application to real-life problems and life-long learning (workforce knowledge)."[4]

As indicated above, the PISA test is administrated every three years, and the latest available data are from the test administered in 2009. The following are the results for the top 10 countries in each of the three areas studied:

- The U.S. is not in the top 10 countries in math, science, or reading.
- Shanghai-China, Hong Kong-China, Singapore, South Korea, Finland, Japan, and Canada are in the top 10 countries in all three disciplines.

The U.S. is basically not competitive on the educational world stage. Notice the dominance of the Asian countries.

The PISA test, while covering all three areas every three years, focuses on one of the three every three years: reading in 2000, mathematics in 2003, science in 2006, and reading again in 2009. The rankings of U.S. students among the top 30 countries in each discipline at the time they were the focus and the results in all disciplines in 2009 are shown in Figure 8-1[5].

Figure 8-1

Rank of
U.S. Students
in top 30 Countries

	Reading	Math	Science
2000	15		
2003		24	
2006			21
2009	17	30	23

The following are the conclusions:

- As far back as 2000, the performance of U.S. students, although measured at different points, has been average or below- average in all three disciplines
- By 2009, the performance had deteriorated in all three major disciplines since the year in which the discipline was a focus.
- In 2009, the performance was in the bottom half of the 30 countries in the league tables, and the performance in math was dead last

In an effort to give you an idea of which countries are ahead of us in the league tables, Figure 8-2[6] shows the actual country rankings in 2009 in mathematics, for example.

Figure 8-2

2009 Rank of U.S. Students in Mathematics

Rank	Country	Rank	Country
1	Shanghai - China	16	Germany
2	Singapore	17	Estonia
3	Hong Kong - China	18	Iceland
4	South Korea	19	Denmark
5	Taiwan	20	Slovenia
6	Finland	21	Norway
7	Liechtenstein	22	France
8	Switzerland	23	Slovakia
9	Japan	24	Austria
10	Canada	25	Poland
11	Netherlands	26	Sweden
12	Macau - China	27	Czech Republic
13	New Zealand	28	UK
14	Belgium	29	Hungary
15	Australia	**30**	**United States**

As you can see, the performance of U.S. students is dismal and embarrassing. However, this is not the worst part. The league rankings for this country relative to the rest of the world condemn us to losing our competitive edge as a country, if we don't do something to change them.

What this means is that over time the growth rate of the US will decline relative to the growth rate for many other countries, and our relative standard of living will decline in a commensurate way. How can we keep up if students in other countries are much better educated that our students are?

This is not something which is being done to us. This is something we are doing to ourselves, and that makes it tragic. We are failing to achieve the second of the two educational objectives which I outlined above, which is international competitiveness.

This situation is unacceptable, and I have some suggestions later in the chapter about how to change it.

The other test I mentioned above is TIMSS, which, as indicated above, measures different aspects of educational achievement. In particular, this test measures the international performance of 4th and 8th graders every 4 years. The latest data for the US, shown in Figure 8-3, are from the test in 2007[7].

Figure 8-3

2007 TIMSS Scores

Area	Score	Average	Rank
4th Grade			
Math	529	500	11
Science	539	500	8
8th Grade			
Math	508	500	9
Science	520	500	11

As you can see, the results are better than the results for the PISA test. Furthermore, U.S. students showed better-than-average improvement from 1995, when the first TIMSS test was administered, to the latest test in 2007

What accounts for the differences between PISA and TIMSS scores?

Let's review the differences again:

- "PISA is the U. S. source for internationally comparative information on the mathematics and science literacy of students in the upper grades at an age that, for most countries, is near the end of compulsory schooling. The objective of PISA is to measure the "yield" of educational systems, or the skills and competencies students have acquired and can apply in these subjects to real-world contexts by age 15. The literacy concept emphasized the mastery of processes, understanding of concepts, and application of knowledge in various situations within subject matter domains. By focusing on literacy, PISA draws not only from school curricula but also from learning that may occur outside of school."[8]
- "TIMSS is the U.S. source for internationally comparative information on mathematics and science achievement in the primary and middle grades. Like NAEP (National Assessment of Education Progress, another testing approach) TIMSS assessments are based on collaboratively developed frameworks for the topics from curricula in mathematics and science to be assessed: but unlike NAEP, the framework and related consensus process involves content experts, education professionals, and measurement specialists from many different countries."[9]
- PISA covers more countries and students than does TIMSS. PISA 2006 covered 57 countries, including all 30 OECD countries, while TIMSS 2007 included 48 countries, including 17 from the OECD.

The different testing approaches and the statistics are complicated and confusing. I have attempted to summarize two different approaches which are broadly used to determine international student achievement.

What does all this mean? PISA seems to be the better approach for determining the qualifications of U.S. students to make their way in the world and for evaluating how these students compare to their counterparts in other countries. If you don't believe this, you could do something simple such as averaging the scores across time frames, countries covered, and, most importantly, different philosophical approaches.

Even with this approach, the conclusions outlined above from the PISA test would still apply. We are not achieving the second educational objective, and this is an extraordinary disservice to our students.

One of the important parts of achieving the first objective of producing educated people is to effectively teach history, both world and American, so that students don't fall victim to the maxim of Santayana, the Spanish philosopher, who said that "those who are ignorant of history are doomed to repeat it". There is no excuse for repeating the mistakes of the past.

How are American students doing in terms of history education? The U.S. Department of Education in 2010 administered a history exam, which is administered every four years, to 7,000 fourth-graders, 11,800 eighth-graders, and 12,400 high school seniors in both public and private schools. The scores are broken into "below basic", "basic", "proficient", and "advanced". Based on the results from the latest National Assessment of Educational Progress (the test to which reference was made above) released on 6/14/11[10]:

- "...only 35% of fourth-graders knew the purpose of the Declaration of Independence"
- "Only 20% of fourth-graders and 17% of eighth-graders who took the 2010 history exam were "proficient" or "advanced", unchanged since the test was last administered in 2006." Proficient means that the students have a solid understanding of the material.
- In high school, only "...12% of 12th-graders were proficient, unchanged from 2006. More than half of seniors posted scores at the lowest achievement level, [which is described as] 'below basic".
- "Fewer than a quarter of American 12th-graders knew China was North Korea's ally during the Korean war..."

These results are abysmal, and they speak to the low level of priority to which the study of history has fallen in school curricula.

The Pulitzer Prize-winning author and respected historian, David McCullough (Truman, John Adams) is very worried about this situation. "We're raising people who are, by and large, historically illiterate...What's more, many textbooks have become 'so politically correct as to be comic'. Having lectured at more than 100 colleges and universities over the past 25 years, he says, 'I know how much these young people - even at the most esteemed institutions of higher learning – don't know...It's shocking"[11].

Two of the more important subjects of the study of history are obviously the Constitution itself and the details of the government (civics – how the government is organized and how it operates) which it created. Where do we stand on these two critical areas?

"Although it is established in the secondary school curriculum, education on the Constitution has suffered from neglect and routine treatment...The educational agenda is cluttered, and priorities often are unclear. In many schools, goals for study of the Constitution may be viewed as no more important than a vast array of competing purposes of education in the social studies...During the 1960s and 1970s, coverage of social history expanded at the expense of political history (including constitutional history)"[12]

There is additional evidence. The Intercollegiate Studies Institute has administered a 60-question multiple choice test to college freshman designed to determine their knowledge of history and our institutions. "Half of the 14,000 incoming freshmen tested failed the 60-question multiple choice test, getting just half the questions right. Worse, they barely know any more when they graduate, with seniors scoring 54 percent correct. No school, not even Harvard or Yale, got above a 69% average among seniors."[13]

These results provide an echo of Santayana's warning. "Colleges like to pride themselves on preparing their young citizens to become future leaders of the Republic, but how can you be an effective leader if you don't know the story of how our nation's past leaders grappled with the perennial challenges of governing a free people?"[14]

History is not the only problem. The same National Assessment of Educational Progress mentioned above has just released (7/19/11) the results for the test it administered in January-March of 2010 to measure proficiency in geography, and the results are no better than the results for their testing of history.

"Only 23% of fourth-graders, 30% of eighth-graders, and 21% of 12th-graders knew enough to be considered "proficient" or "advanced" on the national exam. Proficient means students have a solid understanding of challenging material...only a third of American fourth-graders could determine distance on a map, and less than half of eighth-graders knew that Islam originated in what is now Saudi Arabia..."[15]

It is clear from all that has been outlined above that US students are not competitive with their counterparts around the world, and the consequences are predictable, as I indicated. One explanation might be, some would say, that the US is not competitive in terms of spending on education. It is possible that you get what you pay for and that we are not paying enough.

However, the truth is dramatically different. As you can see in Figure 8-4, which is based on data for 2008, "... with the exception of Switzerland, the United States spends more than any other country on K-12 education, an average of $91,700 per student between the ages of six and fifteen".[16]

Figure 8-4

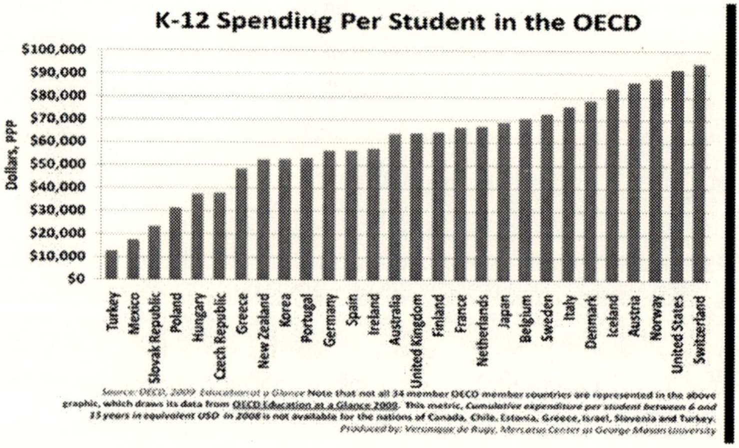

K-12 Spending Per Student in the OECD

The availability of the relative cost data for education in these countries enables us to consider the relationship between: (1) achievement and (2) the cost incurred to produce that achievement. Here are some of observations which can be made:

- Switzerland, which spends the most on education, generally has student performance in the top ten of league tables. This might lead one to conclude that there is a correlation between the amount of money spent on education by a country and the relative performance of its students, However, examples of different relationships dominate the picture.
- The US spends about 1/3 more than Finland, which consistently ranks near the top of the league tables in science, reading, and math.
- The US spends almost 70% more than New Zealand on education, yet in the latest PISA rankings for mathematics, New Zealand ranked #13, compared to the US ranking of #30.
- Norway spends almost as much per student as the US does, but its ranking in math is much better at #21.

The bottom line is that the US spends more on education than any other country in the OECD, and yet its rankings in the league tables are either in the bottom half or actually at the bottom.

The educational establishment in general and teachers' unions in particular offer the following reasons for the poor competitive performance of US students on a global basis: (1) we have a diverse population, which makes it difficult to impose the one-size-fits-all approach which characterizes education in some other countries, (2) we spend a lot of money providing education to special needs children, and (3) we are required to teach a much more diversified curriculum than is the case in many other countries, which detracts from the focus on the specific skills which are tested in standardized tests.

I must say that to me these explanations sound more like excuses. If the objective is to produce students who perform well on a global basis on standardized tests, these issues have to be addressed and not used as rationalizations for poor performance.

Some people would consider the provision of education as a service and not a product. However, I argue that education is a product, not a service, which has to compete in world markets just like any other product. If this product, which has extraordinarily high cost and very poor quality, were being manufactured by a company, the company would have been bankrupt long before now.

President Obama has called for even more spending on education in math and science to close the gap, as the US once did in response to the launch of Sputnik by the Soviet Union 50 years ago. Does this make sense in view of the low productivity we are getting with the money we are currently spending?

"... throwing more money at poorly performing schools has not moved the needle on performance. During the last 40 years, the federal government has spent $1.8 trillion on education, and spending per pupil has tripled in **real** (emphasis provided) terms. Despite the dramatic increase in spending, there has been no notable change in student outcomes."[17]

Using data provided by the Cato Institute, Figure 8-5[18] shows National Assessment of Educational Progress scores in reading, math, and science, along with per pupil spending over the period 1970-2008.

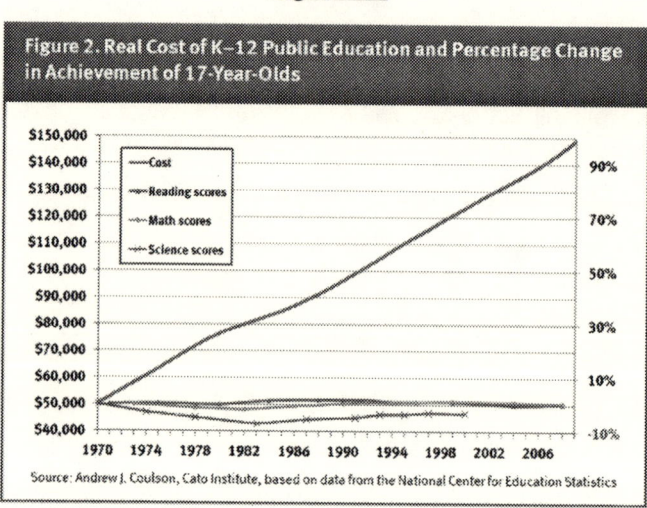

Figure 8-5

Figure 2. Real Cost of K–12 Public Education and Percentage Change in Achievement of 17-Year-Olds

Source: Andrew J. Coulson, Cato Institute, based on data from the National Center for Education Statistics

Note the following;

- During this period, the real cost of K-12 public education increased by almost 100%.
- However, reading, math, and science scores on standardized tests basically did not change. To put it another way, we are now paying almost twice what we paid in 1970 to produce the same results.
- This is a dramatic demonstration of the lack of correlation between spending and achievement. Those who simply push for more spending may have forgotten the famous quote from Albert Einstein – "Insanity: doing the same thing over and over again and expecting different results".

More spending usually means more teachers. As shown in Figure 8- 6[19], the number of students per teacher has fallen steadily since the early 1990s. Ordinarily, this would be considered to be a plus. Presumably, ceteris paribus, smaller class size is better than large class size. However, the problem is that an improvement of 20% in class size in the last 15-20 years has had no impact on the achievement of students.

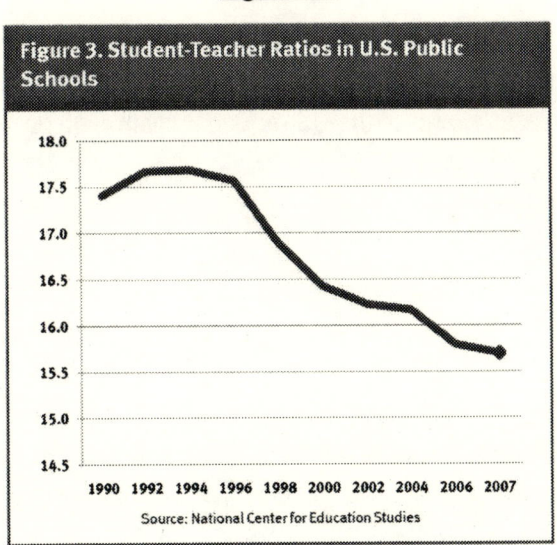

Figure 8-6

Figure 3. Student-Teacher Ratios in U.S. Public Schools

Source: National Center for Education Studies

Another perspective on the competitiveness of US students on a global basis is provided by from the University of Southern California. Based on the data from USC[20], Figure 8-7 shows annual spending per student in each of 12 countries and the country's rank on standardized math and science tests.

Figure 8-7

Country	Annual Spending Per Student ($)	Rank in Standardized Tests - Math	Rank in Standardized Tests - Science
United States	7,743	10	9
United Kingdom	5,834	8	7
Australia	5,766	5	4
Canada	5,749	3	2
Finland	5,653	1	1
France	5,541	7	8
Germany	4,682	6	6
South Korea	3,759	2	5
Japan	3,756	4	3
Mexico	1,975	11	11
Russia	1,850	9	10
Brazil	1,653	12	12

Once again, we can see clearly what the problem is. The US spends more money per student than any of the countries listed (13% more than the UK, which is in second place). However, the US ranking in math is 10 out of 12, and the ranking in science is 9 out of 12. Only Brazil and Mexico have rankings which are worse than those of the US in both categories.

These data confirm the conclusion outlined above – namely, the US education product costs too much and has relatively poor quality in terms of educational attainment.

I would like to add a personal anecdote to the mountain of evidence which demonstrates that our children are not receiving the education they need and which we need them to have to Restore the Future.

About 15 years ago, I was a partner in an investment management company in New York City, and we decided to participate in a program which involved "adopting" a public high school near our office. We had two objectives: (1) we wanted to provide some orientation of the children to an actual office environment by bringing them in, so that they could observe the people who worked there and the work they were doing, and (2) we wanted to do whatever we could do to help improve the academic performance of the students.

This was a school like many others in New York characterized by low achievement and low attendance. It had bars on the lower windows, and to enter the building everyone had to go through a metal detector.

Part of our adoption program was to try to tutor seniors to improve their performance on the SAT. The students with whom we were to work were selected by their teachers, because they felt that these students had the most potential to benefit. The following shows what happened:

- The only students who showed up for the tutoring sessions were girls. Apparently, the boys were either not in school or not qualified.
- We didn't get much response to our invitation to visit our offices, and so we went to the school.
- Only a small number of girls showed up for tutoring (this is a big school), and the students were so far behind in their academic development that there was nothing we could do to get them more prepared for the SAT than they already were at that time.

From the standpoint of educational achievement, these students were already lost, and yet they were supposed to be the best the school had to offer. This, of course, is a sample of only one school, and others in New York may be far better. However, ruining the lives of the children from this one school by not providing them with minimum competence is an outrage, and I hope that each reader feels the same way.

The Future

When a company provides a product which costs too much and which can be demonstrated to be of poor quality, it has only one choice. The quality has to be improved, and the cost has to be reduced. Doing nothing is not an option, unless one desires to settle for educational bankruptcy.

Based on the data shown above, we are already far down the road, but it is not clear that we have yet reached a tipping point. Therefore, it is still possible to turn things around, but doing so will require making improving the quality of education a key national priority like sending a man to the moon and the manned space flight program were when they were launched.

The objectives of this national effort are the ones outlined at the beginning of this chapter: (1) educated men and women, and (2) very high scores on standardized tests. As also pointed out above, education is by definition a responsibility of the states. This makes this effort different from the space program, which was possible to develop by the federal government alone.

"The ... hand of the federal government has its grip on education in the form of the No Child Left Behind federal law on public education, but the trouble is that the law isn't working.'[21] Even Education Secretary Arne Duncan has expressed frustration with the 'slow motion train wreck' that is No Child Left Behind.

In terms of broad perspective, "increasing federal intervention and the resulting burden of complying with federal programs, rules, and regulations has caused a significant growth in state bureaucracy, much of which has a parasitic relationship with federal education programs, straining the time and resources of local schools. Instead of responding first to students, parents, and taxpayers, federal education funding has encouraged state education systems and local school districts to orient their focus to the demands of Washington."[22]

"For more than 45 years, Washington has tried and failed to reform education. Academic achievement languishes, graduation rates have stagnated, and achievement gaps stubbornly persist. It is time for Washington to hand back the reins to state and local leaders..."[23]

Therefore, the answers to the problem of lack of competitiveness of US education have to be found in the states. On the one hand, we must inspire the states to provide a strong fundamental curriculum and to compete with each other in doing so, and, on the other, we must encourage them to "teach-to-the-test".

A strong fundamental curriculum, as outlined earlier, must include studies in at least the following: mathematics, science, reading, writing, grammar and punctuation, history, civics, geography, and foreign languages, as previously outlined on page. Too many of these disciplines are combined into courses which end up being unfocused and unproductive in imparting specific skills.

When I went to school, for example, there was a standard technique for teaching word recognition, spelling, pronunciation, and context. This involved exposing the students to 20 words a week every week and asking them to learn to pronounce them, spell them, and use them in a sentence. After an entire school year, real progress was made in all these areas.

How many schools today teach these areas in this fundamental way? A variety of "new" techniques have been offered by well-intentioned educators, but have any of them worked as effectively as the methods used when I went to school? Sometimes, there really is not anything new under the sun, and efforts to change this fundamental fact only make things worse.

In addition to a strong fundamental curriculum, we must teach-to-the-test. Educators resent being asked to do this, because they think that they can provide a more rounded education which makes the students feel better about learning. In their minds, this justifies the increased expense of providing it. However, the evidence is clear. This approach has not improved educational achievement from the abysmally low level which was reached by earlier experiments in education.

As long as tests measure basic knowledge and the ability to use this knowledge, there is no better way of cross-evaluating students than standardized tests. How else would one compare students across schools, across states, and across countries? Anything else leaves us in a quagmire of subjective judgments, self-justification, and a critical lack of focus.

For the states to achieve the objectives of educated students who perform in an outstanding way on standardized tests, education has to be exposed to competition. Education is one of the few fields of endeavor (medicine is another) which hasn't been exposed to the benefits of competition.

The obvious and necessary way to increase competition in education is to vigorously promote school choice. Parents must be given the ability to choose from among increasingly competitive alternatives which one they think is best for their child. "A lack of educational choice is a reality for thousands of low- and middle-income children across the country and is a major factor in our nation's mediocre academic achievement levels."[24]

Happily, "...2011 has marked a turning-point for school choice, with a growing number of states implementing options such as vouchers, tuition tax credit programs, online learning, and other innovative school choice options – that offer a better alternative for America's children."[25]

"School choice, which saves taxpayers money and simultaneously offers children a higher quality education, is sweeping the nation. It's an idea whose time has come."[26]

One of the most well-known school choice programs is the D.C. Opportunity Scholarship Program, which provides low-income children with vouchers to attend a private school of their choice. The hypocrisy of limiting this program, which has been engineered by Senator Dick Durban and supported by President Obama, was described in Chapter 2.

However, based on the efforts of John Boehner, Speaker of the House of Representatives, the program has now been revived. Dramatic results have been achieved by students in the program. For children who once attended some of the poorest performing schools in the country, academic achievement has risen, and these students have achieved a graduation rate of 91% from excellent private schools, which have no incentive to inflate their graduation rates.

"Indeed, this [school choice] is the new normal: we are taken aback by the states that haven't implemented some sort of school choice option for families, whether it is tuition tax credits, vouchers, or online learning. With every state vying to outpace the next in providing educational opportunities for children, it's hard to keep up with the proliferation of opportunities."[27]

What specifically was happening at the state level in 2011? "Wisconsin has lifted the cap on the Milwaukee Parental Choice Program, the nation's oldest voucher program. Utah has passed a statewide learning program, which allows children in grades 9-12 to take high school coursework online from public or private providers. Indiana has enacted the largest school voucher program in the country, which will help an estimated 600,000 children attend a private school of their choice. Finally, children in Oklahoma may have the opportunity to benefit from a program like the D.C. Scholarship Program."[28]

Let's take a closer look at the Indiana program. "Governor Mitch Daniels of Indiana has signed into law an education reform plan which is based on expanding school choice, increasing school accountability, improving teacher quality, and limiting the stranglehold collective bargaining has had over local schools. According to Daniels, 'Indiana will no longer incarcerate any family's kid in a school that they don't believe is working'. To this end, the governor has created a voucher program that within three years will be available to approximately 60 percent of Hoosier families".[29]

The point is not that every one of the approaches being adopted by individual states will stand the test of time. What is remarkable is the extent to which states are creating new educational approaches designed to achieve both of the educational objectives outlined above. This is the approach the Founders intended for education, and it is exciting to contemplate that the new ideas being developed by the states will lead to the achievement of a nationwide standard of educational achievement which will once again be the envy of the world.

Charter schools potentially offer an important competitive alternative to conventional public schools. What exactly is a charter school? In general, the following are characteristic of charter schools:

- They are primary or secondary schools.
- They receive public money, but they operate independently.
- They are not subject to some of the rules, regulations, and statutes which apply to other public schools, in exchange for which they are held accountable specifically for achieving the results set forth in the school's charter.
- They are opened and attended by choice.
- They are part of the public school system, and they are not allowed to charge tuition.
- They are allowed to hire those teachers which they think would be most effective in producing the product for which the school is designed at mutually agreeable wages. The teachers do not have to be members of a teachers' union.
- When more parents want their children to go to charter schools than there is space available, the children who attend are chosen by lottery.

While charter schools operate with waivers from the procedural requirements of district public schools, this does not "... mean a school is exempt from the same educational standards set by the State or district. Autonomy can be critically important for creating a school culture that maximizes student motivation by emphasizing high expectation, academic rigor, discipline, and relationships with caring adults. Most teachers, by a 68% to 21% margin, say schools would be better for students if principals and teachers had more control and flexibility about work rules and school duties."[30]

Basically, charter schools have come into being because of poor performance of conventional public schools. "One needs to consider the impact of restrictive collective bargaining agreements that prevent rewarding good teachers and removing ineffective ones, intrusive court interventions, and useless teacher certification laws.[31]

It is true that there is a small number of charter schools (5,043 in 2009) which serve, perhaps, 3% of the school-age population (1,500,000 students). However, "according to a 2009 Education Next survey, the public approves of steady charter growth. Though a sizeable portion of Americans remain undecided, charter supporters outnumber opponents two to one. Among African Americans, those who favor charters outnumber opponents four to one. Even among public school teachers, the percentage who favor charters is 37%, while the percentage who oppose them is 31%.[32]

Most importantly, perhaps, data provided by the Center for Education Reform as of 2009 indicate that the cost per year per pupil in charter schools is $8,001, while the same cost in public schools is 50% higher at $12,018.[33]

There has been considerable debate about the performance of charter schools between enlightened educators who are committed to exploring alternatives to the sorry state of public education in this country, on the one hand, and public school teachers and administrators who are primarily interested in preserving their jobs, on the other. What do the data show?

"The best studies are randomized experiments, the gold standard in both medical and educational research. Stanford University's Caroline Hoxby and Harvard University's Thomas Kane have conducted randomized experiments that compare students who win a charter lottery with those who applied but were not given a seat. Winners and losers can be assumed to equally motivated, because they both tried to go to a charter school.

Hoxby and Kane have found that lottery winners subsequently scored considerably higher on math and reading tests than did applicants who remained in district schools.

In another good study, the RAND Corp. found that charter high school graduation rates and college attendance rates were better than regular district school rates by 15 percentage points and eight percentage points, respectively."[34]

Despite the encouraging data on the effectiveness of charter schools in improving educational achievement, there are limits in many states on how many charter schools will be allowed. Why is this? The only answer can be the political influence of the teachers' unions. These unions are concerned that competition will have a negative impact on their members (fewer jobs, pressure on salaries and benefits), and their concern is justified. That is exactly what you want and expect.

Happily, parents are voting with their feet, because their need for better alternatives for their children is so important. In many states, lotteries have to be used, because the demand is so great, and parents are pressuring state governments, with considerable success, to sharply increase the allowable number of charter schools. Based on the evidence of improved student performance and pressure from parents, many states are sharply increasing the number of charter schools.

Another important alternative for parents is home-schooling. Parents who are exasperated by the lack of accomplishment in public schools in particular are opting to use standard curricula, which are state-qualified, to give their children the kind of education which they can't get anywhere else. It is important to note that at the end of 2009, based on estimates from the Center for Education Reform, 1,500,000 children were being home-schooled. This is obviously the same number as the estimated charter school population, and it gets very little publicity.

If school choice is such a good idea, it should have application in other countries. "To identify the long-term benefits of school choice, Harvard's Martin West and German economist Ludger Woessmann examined the impact of school choice on the performance of 15-year old students in 29 industrialized countries.

They discovered that the greater the competition between the public and private sector, the better all students do in math, science, and reading. Their findings imply that expanding charters to include 50% of all students would eventually raise American students' math scores to be competitive with the highest-scoring countries in the world."[35]

To this point, I have discussed what is going on in the laboratories which the states represent, all of which are designed to improve the education achievement of their students. This is one of the issues of the non-competitive educational product in a global context which the US produces. The other issue is cost.

Let me say at the outset that a good teacher is one of the most valuable people we have in our society. All of us remember the good teachers we had and how they changed our lives. They should be paid well, and they should be encouraged to work in a dynamic and inspiring environment to which they make a major contribution.

If they are good, we should be able to take full advantage of their skill, rather than constraining them with rules and regulations which severely limit the application of that skill. On the other hand, those teachers who are not performing well in the classroom should be terminated if they cannot perform at an inspirational level.

Remember that I am arguing that education in the US is a product and that our product is characterized by low quality and high cost. Can you see why the following aspects of our educational system would lower quality and drive up our costs?

- Tenure – many times teachers are able to be granted tenure after only two years of teaching. What this means is that the entire future cost of this teacher is baked in the cake, and it is extremely difficult to eliminate such a teacher, regardless of his/her performance in the classroom
- Performance – teachers are not evaluated on the basis of their value-added in the classroom. This value-added should be measured relative to the two national educational objectives I outlined above.
- Bargaining with themselves – Teachers' unions are the perfect example of the kind of incestuous, circuitous behavior by public

employee unions which I mentioned in Chapter 2. Most teachers have to belong to unions. For the privilege of membership, the unions charge dues. They use these dues to massively support and elect education officials with whom they negotiate their compensation. These officials are aware of influence of unions in keeping them employed, and they therefore bend over backwards to meet the union demands for increased compensation. Once the unions get the compensation they demand, they get more dues, and the process starts all over again.

- Hierarchy – Whenever there is a cost-cutting effort which involves reducing the number of teachers in a school, union rules require that the school follow the industrial concept of "last in, first out". What this means is that the teachers with less experience are terminated first, regardless of whether or not they are better teachers.
- Teacher innovation – Teachers' unions work very hard to protect their jobs from competition, and the perfect example is their lack of support for a program like Teach for America. Teach for America is a program which involves recruiting, training, and placing in schools the best and the brightest graduates of major US colleges.

The common complaint from the unions, which demonstrates their insecurity, is that these young teachers are in some way fundamentally disqualified because they don't have proper teaching certifications. Note that the argument is not based on contribution in the classroom.

If you think about what all this means in terms of an industrial analogy, it is equivalent to the following company:

XYZ Corporation is trying to compete in a global market for widgets, and it was founded on an invention by one of the founders. However, at this point, it is losing share of market rapidly, because it has not been able to make the investment to enable it to maintain the original high quality of its product and its widget now costs much more than the widgets of other companies. Incidentally, all of its employees are required to be members of a union.

This, of course, is an unacceptable situation, and it will inevitably lead to bankruptcy. XYZ is obviously aware of the situation, and it has developed a game plan to attempt to deal with it.

- Since its costs are so high, XYZ would like to downsize its labor force and try to produce more with fewer people. The union leaders are at first not willing to participate in any staff reduction and threaten to strike.
- Management then makes a deal with the union under the terms of which it will be able to lay some employees off, but only those who were hired most recently - that is, in inverse order of seniority. This process, of course, is not related to individual contribution. Management had to agree with this because the union threatened an age discrimination lawsuit if older, more senior employees were included in the layoff.
- Over the years, the total compensation of older employees, as a result of successful union bargaining, has ballooned. Efforts to reduce salaries somewhat and rein in astronomically high benefit costs have been stonewalled by the union.
- In an effort to improve quality and reduce costs, XYZ wants to hire new employees with creative, new ideas. However, the union leaders think that such people are unqualified and force management to: (1) make extensive investments in educating and certifying new employees, and (2) justify every new employee on the basis of potential contribution compared to the existing employee who might be laid off.
- Because of the shortsightedness and unwillingness of union leaders to participate in creating a more competitive product and saving the company, revenue falls dramatically, and XYZ files for bankruptcy.
- At this point, ironically, XYZ has its best chance to succeed as a viable business. The bankruptcy process will protect the company from creditors, while a bankruptcy judge will re-organize the company by reducing employment, eliminating and modifying work rules, reducing costs, and giving the company the operating flexibility to improve the quality of its widget.

The analogy isn't perfect, and the American system of education obviously cannot go bankrupt financially. However, it can go bankrupt conceptually and philosophically, which is what is happening now, and in the process compromise the continued achievement of the American dream by our children and grandchildren.

The bottom line is that the educational product of the US is not competitive, and the educational system in this country must be immediately overhauled. Objectives must be clearly established, and the states must be encouraged to act as laboratories to develop the ideas which will help us achieve these objectives. This is the only constitutional approach which is permitted. The federal government cannot by law and should not, based on the abysmal performance in education to date, be involved in the rescue of education in the US.

Bottom Line

- There should be national objectives for education.
- The appropriate national objectives should be to: (1) provide a strong fundamental education and (2) produce outstanding standardized test scores on a global basis.
- States have the responsibility under the Constitution for education. The Federal government has no constitutional responsibility.
- Based on the results of two different standardized tests, which cover mathematics, science, and reading and which are administered regularly in a broad range of countries, US children are not competitive on the world stage. In one of them, US children are in the bottom half.
- The long-term cost in terms of economic development relative to the rest of the world is enormous.
- The US spends more on education per student than almost any other country
- The low rank on standardized tests and the high rank in terms of cost per student mean that our education "product" is of poor quality and costs too much compared to that of other countries.

- This is because of lack of competition in US education.
- The only way to introduce competition is to vigorously promote "school choice" – parents have to be given the opportunity to vote with their feet.
- Given the lack of focus on the poor state of US education, many different alternatives must be explored by the states, including vouchers, charter schools, tuition tax credits, etc.
- The number of these alternatives is growing rapidly, and evidence suggests that they can provide higher quality education for our children and beneficial competition for conventional schools.

CHAPTER 9
FREE ENTERPRISE

Practically all government attempts to redistribute wealth and income tend to smother productive incentives and lead toward general impoverishment. It is the proper sphere of government to create and enforce a framework of law that prohibits force and fraud. But it must refrain from specific economic interventions. Government's main economic function is to encourage and preserve a free market. When Alexander the Great visited the philosopher Diogenes and asked whether he could do anything for him. Diogenes is said to have replied: "Yes, stand a little less between me and the sun." It is what every citizen is entitled to ask of his government.

Henry Hazlitt, Economist

The great dialectic of our time is not, as anciently and by some still supposed, between capital and labor; it is between economic enterprise and the state.

John Kenneth Galbraith, Economist

So that the record of history is absolutely crystal clear. That there is no alternative way, so far discovered, of improving the lot of the ordinary that can hold a candle to the productive activities that are unleashed by a free enterprise system.

Milton Friedman, Economist

Background

There are so many inadequate definitions of "free enterprise" and so many people who are ill-informed when they speak about free enterprise that it is imperative at the outset to provide an unequivocal definition. One of the best definitions, in my opinion, which includes all the important elements, is the following:

"Conduct of a business without direct government interference; conducting business primarily according to the laws of supply and demand: risking capital for the purpose of making a profit."[1]

- We are talking about a business or enterprise, organized to provide a product or service to individuals who can decide on their own whether or not to purchase them, based on their own experience and needs.
- This is business conducted without direct government interference. The only way it could be otherwise would be to posit that the government would somehow systematically add to the performance of a business. This would have to be because it had more information about what each and every business should do and wise administrators to implement the information across the entire economy. This is a preposterous assumption, without positive historical precedent and with myriad failed attempts.
- The business is conducted according to the laws of supply and demand at prices at which supply and demand tend to be in equilibrium. No one can tell the business better than it can determine on its own how much to produce and sell and at what price.
- The business is created to provide a vehicle for risking capital and making a profit. It follows that without an adequate return on the capital which is put at risk the business cannot survive.

So "free enterprise" refers to businesses which are free from governmental interference in the conduct of their business.

The principal alternative to free enterprise is called the "command and control economy". "Free enterprise and command economies are two opposing economic models that dictate the methods in which economic production and growth should occur within an economy. Free enterprise economies allow individual supply and demand to set price and production. Command economies have their economic production set by the decisions of a central government and may also set the prices of goods for the consumer by the same methods."[2]

These two alternatives have been around since the beginning of time. Free enterprise was what was being implemented in the earliest barter economies and along the earliest trade routes. Command economies were implemented wherever a king or pharaoh existed, who determined how much grain, for example, should be produced and to whom it should be sold.

"The free enterprise system is generally held to be the most efficient method of allocating funds to the most productive entities within an economic system.

For example, under normal conditions, a given area of farmland would be able to maximize productive output under a capitalist [free enterprise] system, which ensures the availability of necessary prerequisites as fertilizer and farm labor. These may not be available under a command economy, because the normal rules of supply and demand are replaced by government decisions. This introduces inefficiencies into the ability of the farmland to produce, so its maximum output would be lower than is possible under freer economic conditions."[3]

The free enterprise alternative is currently the most dominant economic model in the world, precisely because nations and people realize that command economic models don't work. It is certainly true that the free enterprise model is implemented with varying degrees of governmental interference, but most economies are moving in the free enterprise direction.

Think of the former satellites of the Soviet Union. Most of them are trying to implement a free enterprise model overnight, with obviously mixed success. Underdeveloped countries around the globe are trying to determine what resources or products they can sell at world prices determined by free enterprise economies. Even China, the largest economy in the world which is still communist and has elements of a command economy, has a large and rapidly growing free enterprise system, which is one of the marvels of the economic world today.

In 1776, Adam Smith, who was a pioneer in the study of political economy, a social philosopher, and a key member of the Scottish Enlightenment, published what would become one of the most influential books ever published in the fields of business and economics. The book was <u>Wealth of Nations</u>, and it provides the basic philosophical foundation for free enterprise. Coincidentally, it was published in the same year that the Declaration of Independence was signed.

I am going to provide a lengthy quote from the book, because it is so important to understand in the original what Smith is saying [emphasis added]:

"As every individual, therefore, endeavours as much as he can both to employ his capital in the support of domestick industry, and so to direct that industry that its produce may be of the greatest value: every individual necessarily labours to render the annual revenue of the society as great as he can. He generally, indeed, **neither intends to promote the publick interest, nor knows how much he is promoting it.**

By preferring the support of domestick to that of foreign industry, he intends only his own security; and by directing that industry in such a manner as its produce may be of the greatest value, **he intends only his own gain, and he is in this**, as in many others cases, **led by an invisible hand to promote an end which was no part of his intention.**

Nor is it always worse for the society that it was no part of it. **By pursuing his own interest he frequently promotes that of the society more effectually than when he really intends to promote it.** I have never known much good done by those who affected to trade for the publick good."[4]

Smith's famous dictum, which follows from the above, is "It is not from the benevolence of the butcher, the brewer, or the baker, that we expect our dinner, but from their regard to their own interest. We address ourselves, not to their humanity but to their self-love, and never talk to them of our own necessities but of their advantages"[5]

Smith added that "... productive labor should be made even more productive by deepening the division of labor. Deepening the division of labor means under competition lower prices and thereby extended markets. Extended markets and increased production lead to a new step of reorganizing production and inventing new ways of producing which again lower prices, etc., etc. **Smith's central message is therefore that under dynamic competition a growth machine secures "The Wealth of Nations".**[6]

Smith's presentation has been criticized as laissez-faire economics, where anything goes, relying too much on self-interest, without considering society as a whole more directly, promoting greed, and encouraging business to rip off the public. This is far from the case, as he himself recognized.

Smith was not a dreamer with no awareness of the real world. "... he was wary of businessmen and warned of their 'conspiracy against the public or in some other contrivance to raise prices. Again and again, Smith warned of the collusive nature of business interests, which may form cabals or monopolies, fixing the highest price 'which can be squeezed out of the buyers'. Smith also warned that a true laissez-faire economy would quickly become a conspiracy of businesses and industry against consumers, with the former scheming to influence politics and legislation."[7]

The Wealth of Nations has become as influential for the fields of business and commerce as Darwin's Origin of Species was for biology and evolution. The basic disciplines of modern-day industry and commerce have changed little from those outlined by Smith.

The right way to think about the concept that Adam Smith develops in the Wealth of Nations is that it maximizes liberty within the framework of the social contract, as described in Chapter 5. There have to be, as part of the social contract, prohibitions against such activities by businesses, as colluding, the creation of monopoly power, the exploitation of work forces, the use of hazardous materials, and the provision of inadequate consumer safety. However, that framework merely sets the rules within which the "invisible hand" is allowed to operate.

A critical part of the concept is competition. Competition helps allocate resources, determines what products should be sold, determines at what price they should be sold, and drives the innovation which ultimately benefits all consumers. Competition does such a better job of these things than any government ever could. To paraphrase Ronald Reagan, "Competition wins; government attempts to manage the economy lose".

However, there is price to be paid for competition, and nothing can be done to avoid it. Competition and innovation create new industries, new jobs, and new demand for certain skills. Some workers inevitably get left behind, because they did not see change coming and did not prepare themselves adequately for the new world. There is, therefore, a period of adjustment for such workers until they can be re-absorbed into the economy in other jobs.

Joseph Schumpeter, an Austrian-American economist and political scientist, popularized and is most associated with the expression "creative destruction". "The term is used to describe the process of transformation that accompanies radical innovation. In Schumpeter's vision of capitalism, innovative entry by entrepreneurs was the force that sustained long-term economic growth, even as it destroyed the value of established companies and laborers that enjoyed some degree of monopoly power derived from previous technological, organizational, regulatory, and economic paradigms."[8]

Obviously, creative destruction only is justifiable if people buying the newly created products and services are better off after the destruction.

What are examples of "creative destruction"?

- When airplanes came along, they transformed the railroad industry, and the result was a reduction in railway jobs.
- When I was a child, our family used to travel to the Delaware seashore, taking a ferry across the Chesapeake Bay. When the Bay Bridge was built, ferry service was no longer needed, and all the employees associated with the ferry had to find other employment.

- When the integrated circuit was developed, it made consumer products powered by vacuum tubes and diodes obsolete, with a resulting loss of jobs for those employed in making these devices
- When computers were developed, no one needed a slide rule, which was the mainstay of engineering disciplines at the time, any more. Slide rule makers, of course, went out of business.
- "Companies that once revolutionized and dominated new industries – for example, Xerox in copiers or Polaroid in instant photography have seen their profits fall and their dominance vanish as rivals launched improved designs or cut manufacturing costs."[9]

While these examples may seem like ancient history, there are plenty of examples which are taking place all around us today, and some examples are listed below:

- When the internet and email were developed, an alternative to the Post Office was created, and it is still reeling and may not recover from the resulting loss of personal mail. The Post Office continues to reduce staff.
- "Just as older behemoths perceived to be juggernauts by their contemporaries (e.g., Montgomery Ward, FedMart, Woolworths) were eventually undone by nimbler and more innovative competitors, Wal-Mart faces the same threat. Just as the cassette tape replaced the 8-track, only to be replaced in turn by the compact disc, itself undercut by MP3 players, the seemingly dominant Wal-Mart may well find itself an antiquated company of the past. This is the process of creative destruction in its technological manifestation."[10]
- What about the impact that cable news channels and online free newspapers are having on traditional newspapers? Circulation for most newspapers is in a steady decline, and a number of papers, including the Christian Science Monitor and the Seattle Post-Intelligencer, have stopped publishing a traditional newspaper.

In every one of these examples, as is inevitably the case, people are better off in some way or many ways, or in a free market they would not have happened.

People have voted with their feet. They have signaled that they don't want the old and they want the new, and their implicit shift from one alternative to the other is what leads to creative destruction.

"Creative destruction" sounds at first as if it is an undesirable feature of free enterprise. However, in these examples, it can be seen to be an essential part of economic growth. Competition leads to creative destruction which leads to economic growth and improvements in the standard of living.

"competition > creative destruction > economic growth and improved living standards"

"Though a continually innovating economy generates new opportunities for workers to participate in more creative and productive enterprises (provided they can acquire the necessary skills), creative destruction can cause severe hardship in the short term, and in the long term for those who cannot acquire the skills and work experience." [11]

However, in the long run, society as a whole enjoys an improvement in the overall quality of life due to competition and creative destruction, as the above examples demonstrate. Very few people would vote to turn back the clock on any of these developments.

As indicated above, the free enterprise system is the greatest engine for economic growth which the world has ever seen. Unfortunately, with over-regulation, high taxes, costs arbitrarily imposed by the government, and lack of understanding about what activities really produce jobs, government is jamming a stick into the gears of the engine, and, not surprisingly, performance deteriorates.

This is what is happening with the Obama administration, and the unfortunate thing is that they don't even realize it. Is it any wonder that job creation in this economic recovery is the most anemic for any recovery in 50 years?

If the free enterprise system is the greatest engine for economic growth, then it should be possible to demonstrate that it works as advertised. There must be a positive relationship (formally, a correlation) between economic freedom and economic growth. Presumably, the more economic freedom, the faster the economic growth and the greater the achievement of higher standards of living.

There are a number of ways to demonstrate that this relationship exists, and I show below three of them.

Figure 9-1[12] shows the relationship between: (1) the Economic Freedom of a country, based on the Index of Economic Freedom created by the Heritage Foundation and the Wall Street Journal, and (2) that country's gross domestic product per capita. Gross domestic product per capita is the value of all the goods and services produced in the country divided by the number of people in the country. The more goods and services produced per person, the higher the national wealth.

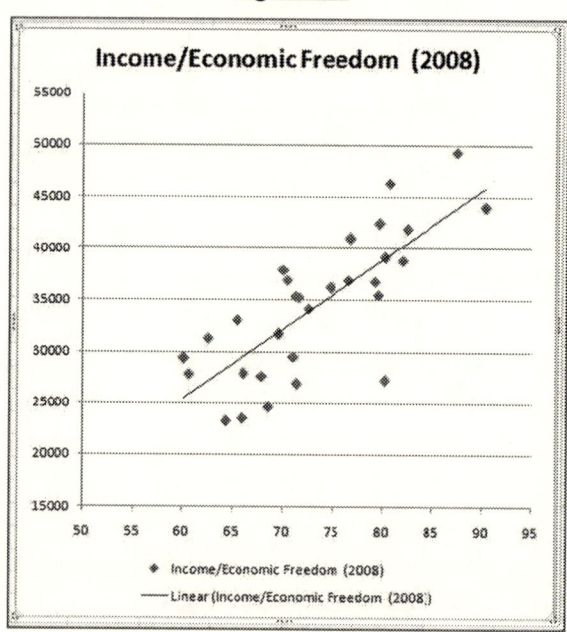

Figure 9-1

Each country is represented by a dot which shows the income (gross domestic product) per person and its level of Economic Freedom.

The important conclusion from Figure 9-1 is that there is a very positive relationship between income per person and economic freedom. That means that, <u>ceteris paribus,</u> the more economic freedom within the social contract the "richer" its citizens will be. Given the strength of this relationship, it is a mystery why every country is not trying to maximize economic freedom.

I pointed out in Chapter 2 that the US is starting to decline in terms of its economic freedom for the first time in memory. You can see the significance of this development in Figure 9-1. If this decline continues, our gross domestic product per person will decline as well, and we will be overtaken by counties with more economic freedom and the higher gross domestic product which follows.

One of the pillars of the Second American Revolution has to be to make sure that this does not happen.

Another way to consider the relationship between economic freedom and standard of living is shown in Figure 9-2[13]. It is clear that the same relationship exists as the one in Figure 9-1, but in this case, the data are aggregated by area of the world, and the top five countries are compared against the bottom five.

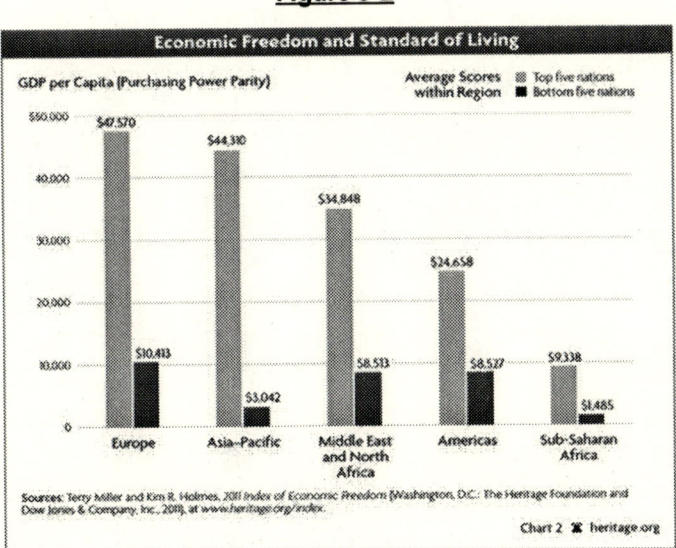

Figure 9-2

Economic Freedom and Standard of Living

Region	Top five nations	Bottom five nations
Europe	$47,570	$10,413
Asia-Pacific	$44,310	$3,042
Middle East and North Africa	$34,848	$8,513
Americas	$24,658	$8,527
Sub-Saharan Africa	$9,338	$1,485

Sources: Terry Miller and Kim R. Holmes, 2011 Index of Economic Freedom (Washington, D.C.: The Heritage Foundation and Dow Jones & Company, Inc., 2011), at www.heritage.org/index.

Chart 2 — heritage.org

In every area of the world, those nations with the highest level of economic freedom have the highest standard of living, and the ratio of improvement from the top five to the bottom five is on the order of 3-14 times. It is important to note, therefore, that free enterprise, which is a significant portion of economic freedom, has an effect which is not determined by geography or stage of development.

However, there are those who don't think that GDP per capita is the best measure of well-being. While that may be true, it is not necessary to debate this question. As you can see in Figure 9-3[14], there is a strong correlation between economic freedom and well-being, as defined by the 2010 Legatum Prosperity Index.

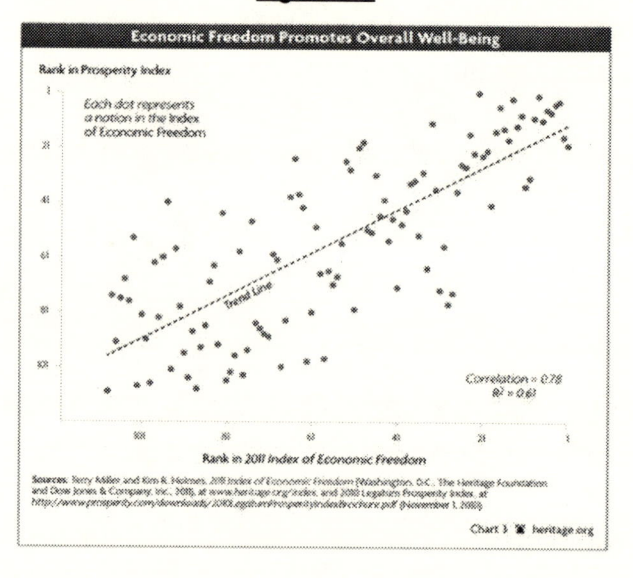

Figure 9-3

"Most people would intuitively agree that 'prosperity' is not just about money but also the quality of life. The Index defines prosperity as both wealth and wellbeing, and finds that the most prosperous nations in the world are not necessarily those that have only a high GDP, but are those that also have happy, healthy, and free citizens."[15]

"The Prosperity Index assesses 110 countries, accounting for over 90 percent of the world's population, and is based on 89 different variables, each of which has a demonstrated effect on economic growth or on personal wellbeing. The Index consists of eight sub-indexes, each of which represents a fundamental aspect of prosperity."[16]

Based on the assessment, countries are given a score and ranked. The top 10 countries in order are Norway, Denmark, Finland, Australia, New Zealand, Sweden, Canada, Switzerland, Netherlands, and the United States.

It is clear that the same general relationship which is shown in Figure 9-1 exists in Figure 9-3, even though the measurement has changed from GDP per capita to "well-being"

Figure 9-4
Freedom vs. Well-Being

Figure 9-5
Freedom vs. Well-Being

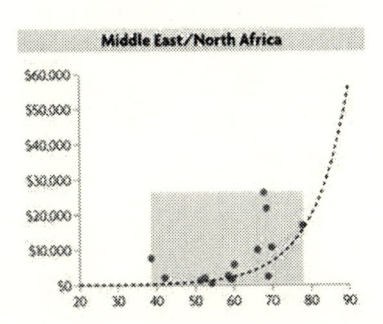

Figure 9-4[17] and Figure 9-5[18], which are part of the overall data set presented in Figure 9-3, demonstrate a very important point. The strong relationship between economic growth and well-being, which characterizes the world as a whole, applies in such disparate areas as Europe and MiddleEast/North Africa. The former obviously consists of developed economies, while the latter consists of developing economies. Regardless of the state of economic development, the relationship in Figure 9-3 applies.

Finally, some people would say that while it is possible that the very positive relationship which exists around the world between economic freedom and

GDP per capita and between economic freedom and well-being, there is still a major problem.

All of this economic freedom and economic advancement is creating severe damage to our environment.

Figure 9-6[19] show the relationship between economic freedom and environmental performance. The environmental data are provided by the 2010 Environmental Performance Index, the sources for which include the Yale Center for Environmental Law and Policy and the World Economic Forum.

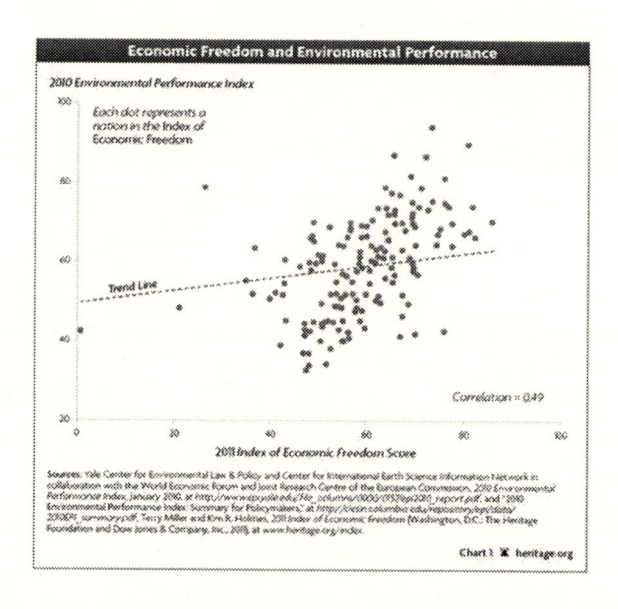

Figure 9-6

Once again, there is a positive relationship between economic freedom and environmental performance, although it is not as strong as the relationships shown in Figures 9-1, 9-2, and 9-3. At least, it is not the negative relationship which some might expect. Once again the relationship is consistent across areas of the world, with the developed countries having higher freedom scores and environmental scores and the developing countries having scores which are lower in both cases.

"... the Index of Economic Freedom strongly suggests that [a] command-and-control approach to "going green" is a fundamentally misguided one. It is the nations whose economies are ranked as most free that do the best to protect the environment, while the least free ones do the worst. Thus, the same free-market principles that have proven to be the key to economic success can also deliver environmental success and point the way to an approach that advances both concerns."[20]

How can this be? The reason seems to be that countries with low per-capita income progressively degrade the environment up to a point. Then, as per capita income keeps increasing, a turning point is reached where the environmental impact starts to improve.

This is shown in Figure 9-7[21], which is what is called a Kuznets Curve.

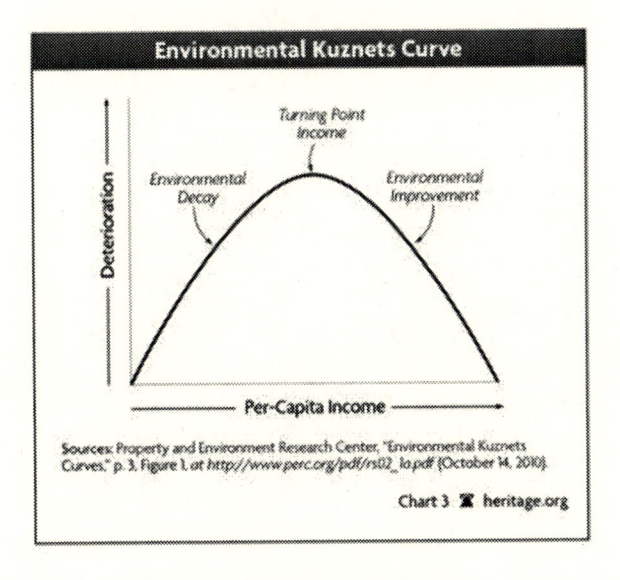

Figure 9-7

"The exact level of wealth needed before things start to become cleaner varies across countries and among different environmental concerns, but the general trend is clear."[22]

Whether we are talking about: (1) GDP per capita related to economic growth and free enterprise, (2) well-being more broadly defined, or (3) the environment, it is clear that free enterprise maximizes liberty, which is one of the principal responsibilities of government, as discussed in Chapter 5. No economic system every invented comes close to matching the desirable results which free enterprise creates.

The problem is that many people, including most Progressives, feel that this is not enough. There are things which the government must do to make things better. In general, as discussed in Chapter 5, governmental interference reduces liberty. However, there is quantitative information about this relationship as well. Consider Figure 9-8.[23]

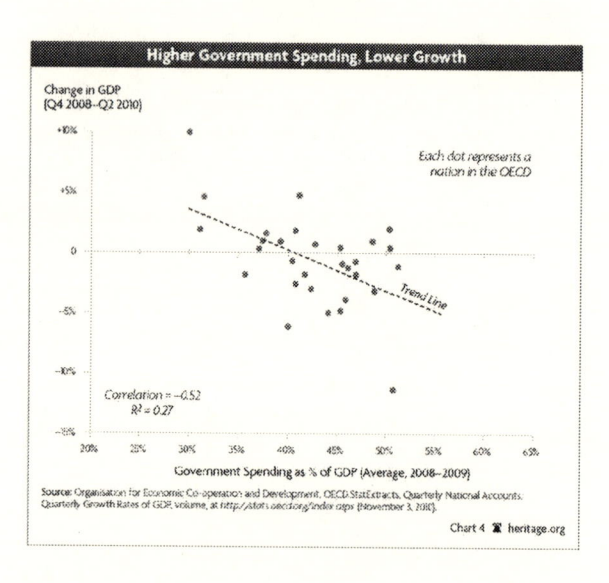

Figure 9-8

In Figure 9-8, the change in GDP growth over the period from Q4:2008 through Q2:2010 is compared to government spending for each of the countries in the OECD. The downward sloping line indicates that, during this period at least, there was a negative relationship between economic growth and government spending as a % of GDP.

This, of course, is exactly what you would expect. If, as described above, free enterprise has its most positive effect in countries which are most free and in which governments are most promoting liberty, then it would follow that these effects would be muted or reversed by increasing government involvement in the economy. Increasing government involvement in the economy for whatever reason by definition reduces liberty.

While I am presenting the evidence which shows the amazing impact of free enterprise, I am not unaware of the problems which can be created by creative destruction. I think that there should be a safety net – this is one of the things which government can do to preserve liberty. However, it must be a safety net which by definition lasts a short time and does not promote long-term dependency.

Free enterprise promotes economic growth, well-being for the citizens, and the environment, and it maximizes liberty. This is as much or more than any reasonable person could expect from an economic system.

"The diversity of the world's peoples and cultures implies that there will be many paths to economic development and prosperity. Indeed, the fundamental value that underpins an economically free and open society is the empowerment of individuals to choose their own paths to prosperity. The whole idea of economic freedom is to increase opportunities of diverse types of activities, with free and open markets as the ultimate arbiter of societies' values and desires."[24]

The free enterprise engine functions best with minimal governmental interference. Most of what government does, as mentioned above, involves sticking things into the engine, which, of course, have the effect of slowing it down. Every citizen should examine each proposed governmental action and determine whether it sticks unwanted things into the engine. Most of the time, it will, and it should therefore be rejected outright on the basis that it does not promote liberty.

This is the negative side of the story. On the positive side, government can do a variety of things, such as reducing regulation, cutting taxes, and promoting free markets and free trade. As we see today, the free enterprise engine is sputtering, and the reasons are now obvious. So is the solution.

As the person who co-founded with me the institutional investment management business with which I was involved for 40 years likes to say, "If you are not moving forward, you are moving backward."

Free Trade

Free trade is the natural extension of domestic free enterprise, and therefore it is appropriate to understand how it works and what its advantages are.

From the 16th to the late-18th century, economic interaction between countries was dominated by "mercantilism". "Mercantilism is the economic doctrine that says government control of foreign trade is of paramount importance for ensuring the prosperity and security of a state. In particular, it demands a balance of trade."[25] It was characterized by high tariffs, export subsidies, monopolizing markets, maximizing the use of domestic resources, exclusive trade with colonies, etc. It was the antithesis of free trade.

"In the English-speaking world, its ideas were dealt intellectual [and ultimately fatal] blows by Adam Smith in the Wealth of Nations and later David Ricardo with his explanation of comparative advantage..."[26]

One of the major objectives of Adam Smith in that book was to demonstrate the falsity of mercantilism. Smith pointed out that if one country (country A) can produce a product, such as computers, at much lower cost than another country, but country B can produce a product, such as t-shirts, at much lower cost than country A, then it follows that Country A should trade with Country B. In such a case, each country takes advantage of its particular efficiency in production.

However, "In the language of some modern economists, an economic exchange is a "zero-sum game", in which there is a winner and a loser. Thus, when Britain trades with France, if one gains by this exchange, the other must lose."[27]

Adam Smith's argument is that economic exchanges are not zero-sum games and that both parties can gain from the exchange. This insight provided a very powerful support for free enterprise in the western world

"In the language of economics, this became known as the absolute advantage argument for foreign trade. This argument, moreover, is not limited to international trade. It applies to trade within a country as well"[28]

"Americans should appreciate the benefits of free trade more than most people, for we inhabit the greatest free-trade zone in the world. The fifty states trade freely with one another, and that helps them all enjoy great prosperity. Indeed, one reason why the United States did so much better economically than Europe for more than two centuries is that America had free movement of goods and services while the European countries "protected" themselves from their neighbors."[29]

"The proper governmental policy toward international trade, Smith held, should be the same as that toward domestic trade – one of letting voluntary exchanges take place in free unregulated markets."[30]

The final nail in the coffin of mercantilism was provided by the English economist, David Ricardo, who published his seminal book, On the Principles of Political Economy and Taxation, in 1917. Ricardo addressed an important issue which Smith did not address.

It is one thing to see why trade would be beneficial between two countries each of which has an absolute advantage over the other in the production of certain products.

However, the issue which Ricardo addressed is "What happens when one country has an absolute advantage over another country in everything?" That is, it can make all products more cheaply than the other country. Most people would say that under these circumstances there is no reason for there to be any trade between the two countries.

Ricardo introduced the idea of "comparative advantage". "The law of comparative advantage says that two countries (or other kinds of parties, such as individuals or firms) can both gain from trade, if in the absence of trade, they have different relative costs for producing the same goods"[30]. This means that the answer to the above question is that, in fact, there would be trade.

Consider the example shown in Figure 9-9.

Figure 9-9

BUSHELS PRODUCED IN ONE DAY

Country	Corn	Potatoes	Opportunity Cost	
A	50	150	1C=3P	1P=1/3C
B	5	25	1C=5P	1P=1/5C

- Country A can produce 50 bushels of corn per day **or** 150 bushels of potatoes, compared to Country B which can only produce 5 bushels of corn **or** 25 bushels of potatoes in a day. Clearly, Country A has a comparative advantage in both corn and potatoes.
- The opportunity cost (what would be given up) for Country A to produce a bushel of corn is 3 bushels of potatoes. The opportunity cost for Country B to produce a bushel of corn is 5 bushels of potatoes. On the other hand, to produce one bushel of potatoes, Country A has to give up 1/3 of a bushel of corn production, while Country B only has to give up production of 1/5 of a bushel of corn.

What happens if the two countries agree to trade at a price which is midway between the opportunity costs – say 1C=4P? The answer is shown in Figure 9-10.

Figure 9-10

AVAILABLE BUSHELS
BEFORE AND AFTER TRADING
(EXCHANGE RATE: 1C=4P)

Country		Corn	Potatoes
A	Before	45	**15**
	After	45	**20**
B	Before	1	20
	After	**1.25**	20

- If Country A gives up the production of 5 bushels of corn, it can produce 15 bushels of potatoes, as shown in Figure 9-10. This is the starting point for Country A. However, if it now sells the 5 bushels of corn on the world market at the exchange rate of 1C=4P, instead of converting that production into potato production, it will receive 20 bushels of potatoes. In the first case, it would end up with 15 bushels of potatoes. Instead, if it trades on the open market, it will end up with 20 bushels of potatoes, instead of 15.
- Similarly, Country B could give up the production of 5 bushels of potatoes and produce 1 bushel of corn. This is the starting point for Country B. However, if it now sells the 5 bushels of potatoes on the world market, which it could have converted into corn production, it will receive 1.25 bushels of corn, instead of only 1.

It is necessary to go through an illustration like this to demonstrate that even in the case where Country A is more effective at the production of both corn and potatoes, trade benefits both. Country A ends up with 5 more bushels of potatoes than it otherwise would have, and Country B ends up with 0.25 more bushels of corn than it otherwise would have.

This is, then, the second part of the argument for free trade. Not only does each country benefit when each has an <u>absolute</u> advantage in one product versus the other (Adam Smith), but each country benefits even if one country has a <u>comparative</u> advantage in all products (David Ricardo).

This is why you could say that increasing trade "lifts all boats", and therefore it is something which all governments should promote. The reverse is also true. If the government raises tariff barriers and institutes other policies to protect domestic industry, trade will naturally decline. This decline will hurt not only the country imposing these hurdles to trade but also the country or countries with whom it would have traded.

These are the theoretical arguments in support of free trade. What do the data show about the effect of world trade on the world economy?

Figure 9-11[31] shows the growth in world exports, which are part of world GDP, annually since 2000, with estimates for 2011. The average GDP growth for the period 1990 to 2008 is probably around 3%, as shown in Figure 9-11. The growth in exports over this same period is almost twice as much. This suggests that exports are raising the level of world GDP.

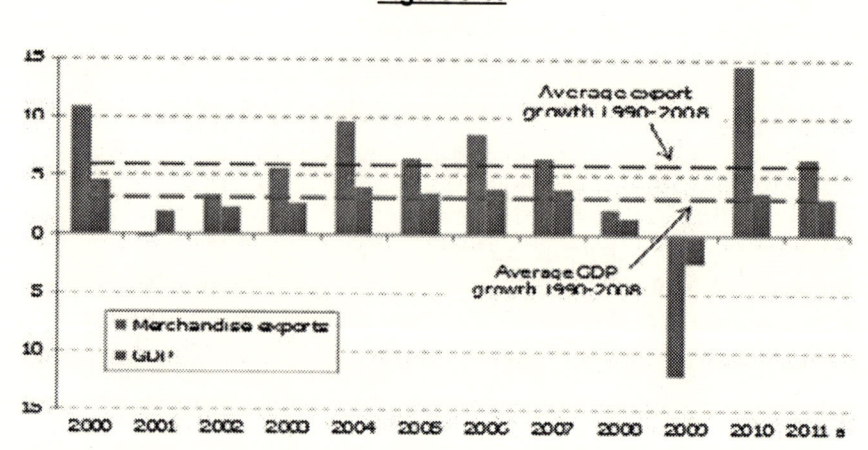

Figure 9-11

Figure 9-12[32] paints a similar picture. In this Figure, the major worldwide trade agreements which basically lowered tariffs on a wide range of products on a global basis and the times they went into effect are shown. The agreements are called "rounds", and they are named after individuals or the location where the talks were initiated.

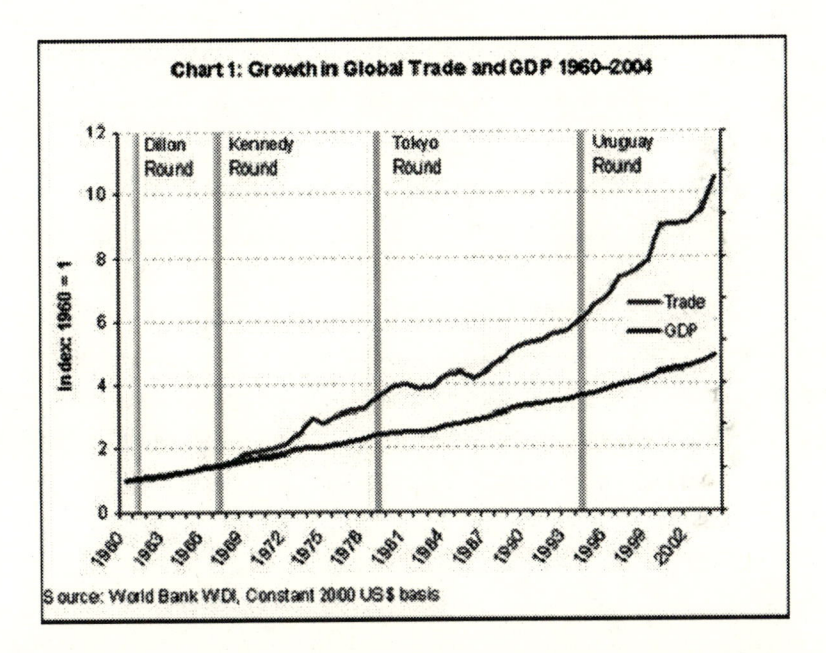

Figure 9-12

Chart 1: Growth in Global Trade and GDP 1960-2004

Source: World Bank WDI, Constant 2000 US$ basis

When a round is finally completed, tariffs are lowered by a large number of countries on a broad range of products in a process designed to promote world trade. As you can see, each completed round has had a positive effect on the growth of world trade

There have not been many rounds, and this is because it takes years to negotiate and finalize a round. An example of this is the Doja round, named after the capital of Qatar in which the first talks took place, started in November 2001. There have been fits and starts, and the round has almost collapsed several times because nations cannot agree on which products to reduce tariffs. The Doja round has thus been going on for ten years, and it has still not been completed.

Not having the Doja round finalized creates an undesirable problem for world trade. Individual countries can be very committed to tariff reduction and feel that it is necessary to pursue their own interests by signing bilateral trade deals. A bilateral trade deal is one between two countries, one country and a trade organization from another part of the world, or two trade organizations. In the process, the tariff reduction process is fractionalized in an inefficient manner.

For example, for a number of years, the US has had bilateral trade agreements pending with South Korea, Columbia, and Panama. These agreements were signed by President Bush, but until recently the Obama administration has not supported their approval by Congress. The principal reason is that labor practices in these countries are not as advanced as those in the US, and it would not support them until the deals are renegotiated to close the gap.

Based on the discussion above, you can see how shortsighted the attitude of the Obama administration can be. It is clear that the US and all nations benefit from free trade. The administration is not allowing US citizens as a whole to avail themselves of the increased array of products at cheaper prices which become available from free trade, because it was standing on ceremony with respect to a relatively small number of workers who might lose their jobs in the process.

This is because it doesn't understand the effect of free trade and the opening up of markets on our trading partners. Trade helps countries develop their economy, and it requires them, to be able to compete in world markets, to modernize their business practices. Without free trade, these practices would be much slower to develop, if they develop at all.

Recently, under pressure from the Republicans and in recognition of the creation of jobs which these agreements would provide, these three agreements have been approved by Congress and signed by the President.

If this attitude were to characterize the approach of the US to trade with the rest of the world, then we would never have started to trade with China. Yet, 35 years after President Nixon made a symbolic trip to China, China is an important trading partner of the US, and one of the most significant countries in the world economic community. Would anyone claim that US consumers as a whole are not better off because we trade with China? Why would the experience with China be any different than the experience with Korea, Columbia, and Panama?

In addition, one can observe that countries which have something at stake – namely, the impact of trade on their economies – are naturally less bellicose, because they have something important to lose. Although it has a strong military and long-standing disputes with neighbors, China now has an enormous stake in the world economic community, and that is a positive for the world.

After the discussion in this chapter, you can see that one of the pillars of the Second American Revolution to Restore the Future has to be vigorous support of free enterprise and free trade.

Bottom Line

- "Free Enterprise" means the ability to conduct a business for the benefit of customers, employees, and shareholders with minimal government interference.
- Free enterprise is the most dynamic engine for economic growth and personal well-being the world has ever seen. Too much government interference causes the engine to operate inefficiently or not at all.
- An individual maximizes the welfare of society by pursuing his own interests with minimal government involvement.
- Economic growth occurs through a process described as "creative destruction", which is a very desirable aspect of economic development.
- There is a strong positive relationship in every area of the world between economic freedom and the resulting economic growth and overall well-being – the greater the former, the greater the latter.
- There is a strong negative relationship all around the world between government spending and economic growth. The greater the former, the less the latter.

- We thus see in the sphere of economics the tradeoff between government and liberty, which was discussed in Chapter 5.
- Free trade promotes economic growth for a country, and it is an important component of free enterprise.

CHAPTER 10
CONCLUSION

And for the support of this Declaration, with a firm reliance of the protection of divine Providence, we mutually pledge to each other our Lives, our fortunes and our sacred Honor.

> Declaration of Independence

I predict future happiness for Americans if they can prevent the government from wasting the labors of the people under the pretense of taking care of them.

> Thomas Jefferson

Let us be sure that those who come after will say of us in our time, that in our time we did everything that could be done. We finished the race; we kept them free; we kept the faith.

> Ronald Reagan

The First American Revolution was both a revolution of ideas and a revolution to overthrow a tyrant militarily. The purpose of the Second American Revolution is to Restore the Future by restoring the ideas and the form of government for which the revolutionaries, and the great military men and women in the years since, have fought and died. Only then can we truly say that the future has been preserved for our children and grandchildren like it was by prior generations.

This Revolution cannot and should not be a military revolution, which would obviously irrevocably rend the fabric of our country. It is not about creating and implementing new ideas. It is about restoring the new ideas from 220 years ago to the pre-eminent position which they once occupied. This Revolution has to be about the ballot box and using the power of the people which the Founders created to effect change.

In every election at every level, the American people must vote for and hold responsible representatives who are committed to constitutional government. The power of the Contract with America which was created in 1994 to effect a sea change in favor of constitutional government can easily be demonstrated.

What we must have now is a Contract with America II, and the candidates who are worthy of your votes must commit to all of the following statements. I will:

- Support the Constitution in its original form and reject the concept of a "living" Constitution which adapts the Constitution to current circumstances, rather than the other way around.
- Support the view of all of the Founders that the natural rights which are enshrined in the Constitution and which have been the foundation of the most remarkable development in the history of civilization come from a superior being. Any other view condemns us to mediocrity and a bias toward some of the other forms of government which have been tried and so spectacularly failed.
- Promote the small central government envisioned by the Founders in the Constitution, with limited intrusions into private lives.
- Support the view that promoting equality of opportunity is vastly superior to promoting equality of condition in creating economic growth, real jobs, and the satisfaction for the American people of knowing that they have created their own destiny. This country is and has been a beacon to people all over the world because we recognize and stress this critical world view.
- Vote for legislation which promotes individual liberty and provides a framework so that everyone can be as successful as he wants and chooses to be.

- Support the rule of law, which is a system of laws based on the authority of the Constitution.
- Promote a strong national defense without which our country and the unique system of government which has made it the envy of the world are doomed. A good defense is a good offense.
- (1) Recognize the lack of competitiveness of American secondary education in terms of relative performance in the world, (2) understand the long-term economic and social cost which this situation will produce, if it is not addressed, and (3) commit to establishing and achieving national education objectives, to be achieved by the states, which will raise our education quality to a level which is among the best in the world.
- Support free market capitalism as the best approach to promoting long-term economic growth and well-being.
- Support free and open trade. In total, a rising tide of international trade lifts all boats. The evidence is clear that standards of living across the globe have benefited to an extraordinary degree because of free trade.
- Create and maintain a system of low taxes for individuals and corporations, because: (1) a dollar spent by government is a dollar taken from the private sector, and (2) many studies have shown conclusively that a dollar spent by the government is in general far less productive than a dollar spent in the private sector.
- Vote in favor of responsible management of the finances of the nation.
- Reduce both the size and the growth rate of entitlements, before we lose the opportunity to recover.

Is this a litmus test? Yes, it certainly is. This is not the time for half-hearted attempts to address and solve these issues. At best, half-hearted attempts will produce half-hearted results, and we cannot tolerate them. Do not let anyone tell you that making and acting on these statements is too much to ask of our elected representatives

These are questions which measure the fidelity of your representatives to the Constitution as it was created and as it has been amended. There is nothing partisan about them, unless a critic would want to paint himself into a corner by proposing that living by the Constitution is "partisan".

Those for whom you vote must be able to answer "yes" to the vast majority of these questions without qualification, and they must be prepared to explain any answer other than an unqualified "yes" to any of them.

If your representatives are unable to answer these questions strongly in the affirmative, they should be voted out of office. They are part of the problem and not part of the solution. To replace them, you must help find and vote for candidates who can answer these questions in the affirmative and help Restore the Future.

Unfortunately, the American electorate is largely apathetic. According to a study by the Center for the Study of the American Electorate at American University, participation in primaries for midterm elections for governor and/or the US Senate by Democrats was just 8.4% in 2006, down from 19.9% in 1962, and by Republicans was just 7.2% in 2006, compared to 12.8% in 1962. The figures are even lower for primaries for the US House of Representatives and state legislatures.

The figures in 1962 were bad enough. The latest figures are an embarrassment. Primaries are an important part of the process, since they are designed to winnow the list of candidates to the best candidates for respective elections. If voters sit on their hands to this extent, they will not end up with the best candidates for each job, and they will get the government they deserve.

At the national level, the average participation in the last five presidential elections is only 53.5% of the voting age population. Having presidential elections determined by only half the electorate is a national disgrace. Citizens have fought and died for 220 years, and are still fighting today, to guarantee the freedom to vote, and yet half the citizenry feels that it can afford not to participate.

The elections in 2012 are more important for the future of our country than any in the last 70 years. These are not elections where the citizens can afford to sit on their hands. You have a right which only a few people in history have had. Don't throw it away, or you will have no right to bemoan or complain about the outcome.

After a century of erosion of the principles of the Constitution, we are reaching a point of no return. The only salvation for the Republic at this point is to get citizens: (1) to participate in their country's political process, as required by the government that the Founders established, and (2) to vote for candidates who meet the litmus test described above. We can accept no less.

These are not political issues, although by definition we can only Restore the Future through the political process. These are not Republican or Democratic issues. They are issues of national survival.

Remember this paraphrase of the quotation from Sun Tzu at the beginning of the Introduction: know your enemy, know yourself, and you cannot fail. After reading this book, you know the enemies of the Republic. If you support the Contract with America II outlined above, you know yourself. Given these two things, our Revolution should be a success.

Please pick up your intellectual arms and become a foot soldier in the Second American Revolution to Restore the Future. As was the case with the first Revolution, the future of our country depends on you.

Restore the Future!

REFERENCES

Chapter 1

1. David Hume, <u>A Treatise of Human Nature</u>, Book III, Part II, Section II: Of the Origin of Justice and Property"
2. "The Collected Writings of Rousseau" edited by C. Kelley and M. Raymond, Volume III, page 361
3. The New American, 11/6/2000, as recorded by Constitution signer James McHenry
4. Freedomsphoenix.com
5. "Jefferson, Education, and the Franchise" , <u>Archiving Early America</u>, as cited in Padover, 1939, page 89
6. Richard Epstein, <u>How Progressives Rewrote the Constitution</u>, pages 2-3
7. David Gordon, Ludwig Von Mises Institute, Spring 2006, Volume 12, Number 1, reviewing Richard Epstein, <u>How the Progressives Rewrote the Constitution.</u>
8. Cato Institute Book Forum summary of Richard Epstein<u>, How Progressives Rewrote the Constitution</u>, Wednesday, 2/15/2006
9. "Glenn Beck, Progressives and Me", Ronald Pestritto, Wall Street Journal, 9/15/10, page A17
10. Ibid
11. Townhall.com, Amanda Carpenter, 7/24/2007
12. George Santayana, the Life of Reason, Volume 1, 1905

Chapter 2

1. "The Morning Bell", Heritage Foundation, 12/10/2009
2. Ibid
3. Heritage Foundation 2009 Annual Report, page 4
4. USA Today, 6/7/11, front page
5. "The Morning Bell", Heritage Foundation, 7/13/11
6. Thomas Sowell, Capitalism Magazine, 12/4/04
7. Thegpproject.com, 6/8/2009
8. Wall Street Journal, 12/8/09, page A20
9. USA Today, 12/11/2009, front page
10. Cato Institute, "Federal Pay Continues Rapid Ascent", 8/24/2009
11. Wall Street Journal, 3/26/10, page A18
12. Ibid
13. Ibid
14. Ibid
15. Ibid
16. Wall Street Journal.com, 1/22/10
17. Wall Street Journal, 1/28/10, page A18
18. Heritage Foundation, "2011 Budget Chart Book"
19. New York Times, 3/25/10
20. Ibid
21. "Why Government Spending Does not Stimulate Growth", Brian Riedl, Heritage Foundation, 1/12/08
22. "The Morning Bell", Heritage Foundation, 4/2/2010, pages 1-2
23. Heritage Foundation, Leadership for America Impact Report, 12/09, page 10
24. "The Morning Bell", Heritage Foundation, 1/20/10

Chapter 3

1. Vice Admiral Bob Scarborough, email 11/27/09
2. ProCon.org, "Under God in the Pledge"
3. Ibid

4. Matthew Spaulding, "We Still Hold These Truths", page 52
5. Ibid
6. Ibid, page 65
7. Ibid, page 52
8. Ibid, page 60-61
9. Ibid, page 56
10. Ibid, page 61
11. Ibid, page 63
12. Ibid, page 63
13. Ibid, page 64
14. Marx on Religion, angelfire.com
15. Ibid
16. Who is Nature's God?, essay by David J. Voelker, 1993
17. Thomas Paine, "The Age of Reason"
18. Albert Henry Smyth, The Writings of Benjamin Franklin (1906 edition), volume IX, page 601
19. Matthew Spaulding, "We Still Hold These Truths", page 154
20. NBC, 5/3/11
21. www.cnsnews.com
22. msnbc.com, 4/15/10
23. Ibid
24. USA Today, 4/27/10, page 1A
25. usatoday.com, 8/10/2005
26. Ibid
27. Ibid
28. nationmaster.com
29. usatoday.com, 8/10/2005
30. Michael Streich, Religion and Secularization in Europe, 5/26/09
31. usatoday.com, 8/10/2005
32. usatoday.com, 5/13/09
33. Ibid
34. Heritage Foundation, Morning Bell, 6/18/10, page 4
35. Ibid
36. Ibid, page 5

37. Wickipedia.org
38. Ibid
39. www.guttmacher.org
40. CIA World Factbook, 2010 www.nationmaster.com
41. John Winthrop, governor of Massachusetts Bay Colony, "A Modell of Christian charity", a discourse written aboard the Arabella during the voyage to Massachusetts, 1630 – Robert C. Winthrop, <u>Life and Letters of John Winthrop</u>, page 19 (1867)

Chapter 4

1. allaboutphilosophy.org/communism
2. 2011 Federal Budget Chartbook, Heritage Foundation
3. Ibid
4. Ibid
5. Ibid
6. Wall Street Journal, 8/17/2011, page A14
7. 2011 Federal Budget Chartbook, Heritage Foundation
8. "The Historical Lessons of Lower Tax Rates", Heritage Foundation, 8/13/03
9. "The Keynesian Growth Discount", <u>Wall Street Journal</u>, 4/29/11. Page A14

Chapter 5

1. "Socrates – Understanding Understanding", RedState Eclectic, Georg Thomas, 6/4/2008
2. Ibid
3. Sanjeev Sabhlok's occasional blog 4/13/11
4. Ibid
5. Ibid
6. www.middle-ages.org/magna-carta
7. Wikipedia.org

8. "Algernon Sidney: the Inspirational Founding Father, Robert A. Ficalora,www.republicandemocracy.us/SydneyHistory
9. Ibid
10. Google, Algernon Sidney, wikiquotes
11. Ibid
12. "Algernon Sidney: the Inspirational Founding Father, Robert A. Ficalora,www.republicandemocracy.us/SydneyHistory
13. Wikipedia, John Locke
14. Ibid
15. Bartleby.com
16. http://everything2.com/title
17. http://thinkexist.com/quotations
18. www,Monticello.org
19. http://everything2.com/title
20. Mises Economics Blog, Gary Galles, 4/23/07
21. "Liberty and the Constitution", Jacob Hornberger, The Future of Freedom Foundation, 8/01
22. http://thinkexist.com/quotations
23. "We Still Hold These Truths", Matthew Spaulding, page 8
24. Ibid
25. Ibid, page 9
26. Ibid
27. Ibid
28. Ibid, page 26
29. Ibid
30. Ibid
31. http://www.firstprinciples.us

Chapter 6

1. "We Still Hold These Truths", Matthew Spaulding, page 83
2. Ibid
3. Ibid, page 84
4. Ibid, pgs. 84-86

5. Ibid, page 86
6. Ibid, page 88
7. Ibid
8. Ibid, page 89
9. Ibid, page 99
10. Ibid, page 113
11. Ibid, page 114
12. Ibid, page 115
13. Ibid, page 116
14. "The Morning Bell", Heritage Foundation, 6/29/11
15. Ibid
16. "Solutions for America" – page 17, Volume 9 – August 17, 2010, "The Rising Tide of Red Tape", Heritage Foundation
17. Ibid
18. Ibid
19. 'Solutions for America' – page 29, Volume 15 – August 17, 2010, "Overcriminalization"", Heritage Foundation
20. "The Morning Bell", Heritage foundation, 5/5/10
21. "Solutions for America" – page 31, Volume 16 – August 17, 2010, "Civil Justice Reform", Heritage Foundation
22. Ibid
23. "The Coming Constitutional Debate", Stephen J, Markham, Imprimus, 4/10/10, Volume 39, Number 4, Hillsdale College

Chapter 7

1. "We Still Hold These Truths", Matthew Spaulding, page 173
2. "The Bill of Rights: Antipathy to Militarism, Jacob G. Hornberger, The Future of Freedom Foundation, September 2004
3. Ibid
4. "The State of the U.S. Military', Heritage Foundation, January 2010, page 1
5. Ibid
6. Ibid

7. Ibid
8. "Morning Bell", Heritage Foundation, 5/3/11
9. "The Decline of U.S. Naval Power, <u>Wall Street Journal</u>, 3/2/11, page A17
10. Ibid
11. Ibid
12. www.meetup.com, 1/13/11
13. http://33-minutes.com
14. "Medium Extended Air Defense System: Continued Funding Needed", Heritage Foundation 2/22/11
15. www.meetup.com, 1/13/11
16. 2011 Federal Budget handbook, Heritage Foundation
17. wickipedia
18. http://publicintelligence.net
19. Ibid
20. "Types of Terrorism", about.com
21. www.emedicinehealth.com/biological_Warfare
22. "Time for America to Get Cyber-Serious", page 1, Heritage Foundation, 5/16/11
23. "Backgrounder", page 2, Heritage Foundation, 1/31/11
24. http://publicintelligence.net
25. "Backgrounder", page 3, Heritage Foundation, 1/31/11
26. http://publicintelligence.net
27. Ibid
28. "Pentagon discloses largest-ever cyber theft", <u>Bloomberg Businessweek,</u> 7/14/11
29. "Time for America to Get Cyber-Serious", page 1, Heritage Foundation, 5/16/11
30. "Terrorism Issues", About.com, 6/24/11
31. IBID
32. "Fukushima: Expert", Michael Bunn, www.reuters.com, 6/23/11
33. "The Gates Farewell Warning, Wall Street Journal, 5/28-29/11
34. Ibid
35. Ibid

36. Ibid
37. "Fifth Annual Address Message", George Washington

Chapter 8

1. "International Test Scores", http://4brevard.com/choice/international-test-scores.htm
2. "Constitutional Requirements Governing American Education", http://education.stateuniversity.com
3. Ibid
4. "Programme for International Student Assessment", wikipedia
5. Ibid
6. Ibid
7. "Trends in International Mathematics and Science Study", Wickipedia
8. "Comparing NAEP, TIMSS, and PISA Results *nces.ed.gov/timss/pdf/naep_timss_pisa_comp.pdf*
9. Ibid
10. Wall Street Journal, page A3, 6/15/11
11. Wall Street Journal, page A13, 6/18-19/2011
12. "Education on the U.S. Constitution, ERIC Digest #39", www.ericdigests.org
13. "Opinion: Colleges Get Failing Grades on Civics" www.aolnews.com
14. Ibid
15. "Don't Know Much About Geography", Wall Street Journal, 7/20/11
16. "Losing the Brains Race", Veronique de Rugy, Reason Magazine, March,2011
17. Ibid
18. Ibid
19. Ibid
20. MAT@USC, 2/8/11
21. "Send Education Dollars and Decision-Making Back Home", Morning Bell, Heritage Foundation, 6/13/11
22. "Reducing the Federal Footprint on Education and Empowering State and Local leaders", Backgrounder, the heritage Foundation, 6/2/11

23. Ibid
24. "School Choice is the New Normal", Morning Bell, Heritage Foundation, 5/31/11
25. Ibid
26. Ibid
27. Ibid
28. "The Most Exiting Two Minutes in School Choice", The Foundry, the Heritage Foundation, 5/7/11
29. "Governor Mitch Daniels Champions Education Reforms", The Foundry, Heritage Foundation, 5/6/11
30. Wikipedia, "Charter School"
31. "Charter Schools and Student Performance", <u>Wall Street Journal</u>, page A23, 3/16/10
32. Ibid
33. "K-12 Facts", Center for Education Reform, www.edreform.com
34. "Charter Schools and Student Performance", <u>Wall Street Journal</u>, page A23, 3/16/10
35. Ibid

Chapter 9

1. "Free enterprise", Answers.com
2. "Free Enterprise Vs. Command Economy", Ellis Davidson, smallbusiness.chron.com
3. Ibid
4. "The Wealth of Nations", Wikipedia
5. Ibid
6. Ibid
7. Ibid
8. "Schumpterian Creative Destruction, Wikipedia
9. Ibid
10. Ibid
11. Ibid
12. "Economic freedom and GDP per capita", www.themoneyillusion.com

13. <u>2011 Index of Economic Freedom</u>, Heritage Foundation
14. Ibid
15. <u>2010 Legatum Prosperity index</u>, Executive Summary, Legatum Institute
16. Ibid
17. <u>2011 Index of Economic Freedom</u>, Heritage Foundation
18. Ibid
19. Ibid
20. Ibid, Chapter 4
21. Ibid
22. Ibid, Chapter 4
23. Ibid
24. Ibid
25. "Mercantilism", Wickipedia
26. Ibid
27. "Adam Smith Free Trade International Trade Theory", www.economictheories.org
28. Ibid
29. "Free Trade", Alan blinder, Library of Economics and Liberty
30. "Comparative Advantage" – Wickipedia
31. "Trade Growth to ease in 2011 but despite 2010 record surge, crisis hangover persists", World Trade Organization press release, 4/7/11
32. "Growth in Global Trade and GDP 1960-2004", World Bank WDI

INDEX

A

AAA credit rating 42
abortion 25, 68, 71–74, 126, 136
absolute advantage 203, 206
act of war 153
Adam Smith 189, 190, 202, 203, 206, 226
adjusted gross income 91
adopting 173
agnostic 63
Alexander Hamilton 12, 109, 110, 127, 141
Alexander the Great 186
Algernon Sydney 106, 118
amendment 9, 12, 13, 20, 21, 25, 54, 58, 125–127, 131, 132, 133–136, 138, 139, 160
arbitrary rule 120
Arms Trade Treaty 21
Articles of Confederation 141, 142
atheist 61, 63, 67
Atlas Shrugged 85, 86
Ayn Rand 85

B

Bacon 107
ballot box 212
bargaining 177, 179, 181, 183
benign neglect 103, 128
Benjamin Franklin 11, 54, 108, 219
Big Bang 62, 63
Bill of Rights 9, 10, 106, 121, 125, 126, 132, 136, 138, 222
bioterrorism 150, 153, 157
births 68, 70, 71, 74
budget 1, 18, 22, 31, 31–33, 35, 38, 39, 41, 71, 83, 84, 85–87, 89, 92, 93, 96, 100, 130, 143, 155, 218, 220, 223
Bush 96, 97, 129, 208

C

Capitol 65, 66
Cato Institute 170, 217, 218
Charles Darwin 63
charter school 178, 179, 180, 185, 225
church 25, 26, 54–56, 68–70, 72, 74
Cicero 1, 3
civics 15, 128, 159, 167, 175, 224
Clarence Thomas 12
class size 171
class warfare 22, 98, 115
common defence 102, 140, 146, 148, 155
Communist Manifesto 57, 78, 79
comparative advantage 202, 204, 206, 226
competitive 24, 113, 145, 159, 161, 163, 168, 169, 176, 178, 181, 183, 184
Congress 18, 22, 25–27, 32, 43–45, 65, 75, 76, 86, 87, 97, 115, 123–126, 128, 130, 133, 144, 160, 208
Congressional Budget Office 18, 33, 35
Constitution 2, 8–14, 16, 18–20, 24–26, 36, 47, 52, 54, 56, 58, 77, 85, 101, 102, 104, 106, 108–111, 114, 120–125, 126–128, 131–138, 140, 142, 143, 156, 159, 160, 167, 184, 212–215, 217, 221, 224

Constitutional Convention 11, 54, 55, 61, 107, 108, 123, 125
Contract with America II 212, 215
conventional warfare 145, 146, 148
corporate tax rate 115
creative destruction 191, 192, 193, 201, 209, 225
Creator 8, 14, 56, 59, 61, 63, 65, 102
cultural relativist 49, 50, 51, 62
cyber terrorism 151, 153, 157
czars 27, 115

D

Daniel Patrick Moynihan 76
Darwin 63, 64, 190
David Hume 5, 217
David McCullough 167
David Ricardo 202, 203, 206
D.C. Opportunity Scholarship Program 176
Declaration of Independence 2, 8, 14, 17, 19, 47, 65, 101, 102, 106, 109, 111, 118, 155, 166, 189, 211
Decline of the American Republic 16
deficits 22, 31, 33, 39, 41, 75, 76, 86, 87, 89, 96, 100, 115
Diogenes 186
Discourses on Government 106, 107
District of Columbia 44
Doja round 207, 208
Douglas 65
due process 123, 134
dysfunctional 76

E

earmarks 43
economic freedom 46, 194–199, 201, 209, 225, 226
economic growth 1, 3, 28, 36, 72, 79, 85, 86, 94, 96–99, 97, 191, 193, 194, 197, 200, 201, 209, 210, 212, 213

education 3, 15, 26, 37, 44, 45, 52, 55, 58, 78, 82, 102, 104, 128, 132, 158–161, 165–169, 170–172, 174–182, 184, 185, 213, 217, 224, 225
educational establishment 169
Edwin Meese 120
Efficient 37, 188
elections 11, 76, 86, 94, 128, 214
employment and compensation 28, 47
engine 3, 193, 201, 202, 209
England 23, 25, 106, 122
English common law 122
entitlements 26, 27, 33, 34, 80, 84, 85, 93, 100, 147, 148, 156, 157, 213
enumerated powers 9, 10, 23, 36
environmental 21, 129, 198, 199
EPA 27, 129
equality of condition 23, 75, 78–80, 82, 85, 93, 99, 100, 116, 134, 212
equality of opportunity 23, 75, 78–80, 82, 93, 99, 100, 116, 212
evolution 9, 58, 60, 62–64, 114, 190
exceptionalism 113, 117, 119
exports 206

F

fair share 22, 27, 90–92
Federal civilian 30
Federalist 9, 54, 55, 105, 110, 123, 126, 127, 140, 141
fidelity 15, 120, 121, 213
First Amendment 25, 160
fiscal mismanagement 39
fiscal soundness 86, 88, 100
Founders 3, 8–14, 20–25, 36, 47, 54–61, 67, 70, 74, 77, 82, 83, 99, 103–105, 107–111, 113, 114, 120, 122, 125–127, 131, 133, 137, 141, 142, 148, 177, 182, 212, 215
Founding Fathers 2, 12, 16, 20, 27, 57, 67, 69, 103, 106, 110, 132, 160
Frank J. Goodnow 13

Frederic Bastiat 110
freedom 3, 6, 10, 14, 19, 20, 26, 46, 60, 63, 66, 76, 101, 104, 110–113, 112, 114, 119, 126, 142, 147, 151, 194–199, 201, 209, 214, 221, 222, 225, 226
freedom of religion 10, 60, 63
freedom of speech 10
freedom of the press 10
free enterprise 3, 52, 90, 99, 186–189, 193, 196, 200–203, 209, 210, 225
free trade 3, 52, 116, 144, 202, 203, 206, 208–210, 213, 226
fundamental curriculum 175

G

GDP 34, 38, 40, 41, 87–89, 93, 94, 98, 100, 117, 143, 147, 148, 154–156, 196–198, 200, 206, 225, 226
General Motors 80
geography 159, 168, 175, 196, 224
George Santayana 15, 217
George Washington 56, 113, 127, 141, 156, 224
Gerald Ford 75
Glorious Revolution 122
God 14, 49, 54–56, 58–61, 59, 64, 65, 66, 67, 70, 74, 102, 107, 218, 219
Great Depression 115
Greek 59, 103, 150
grievances 2, 17–20, 28, 46, 47, 155

H

Harry Truman 42, 101
Henry David Thoreau 5
Henry Hazlitt 186
Heritage Foundation 18, 46, 54, 71, 97, 145, 194, 218–220, 222, 223–226
high cost 170, 181
Hillsdale College 120, 131, 134, 222

history 1, 6, 9, 10, 13–15, 24, 42, 58, 64, 66, 68, 77, 93–95, 99, 103, 108, 110, 114, 115, 117, 121, 125, 127, 131, 133, 134, 136, 141, 158, 159, 166–168, 175, 186, 192, 212, 215
home-schooling 180
Hugo Black 25
hyphenated American 117
hypocrisy 28, 43, 44, 47, 176

I

illiterate 167
immigration policy 117
income tax 14, 78, 80, 96
Index of Economic Freedom 46, 194, 199, 226
inheritance 78, 80
intellectual bankruptcy 78
interstate commerce 24, 25
invisible hand 189, 190

J

Jack Kemp 96
James Madison 9, 105, 110, 113, 125, 140
James Wilson 109
Jean Jacques Rousseau 6, 112, 217
jobs 22, 28, 36, 94, 96–98, 115, 116, 129, 179, 180, 182, 191–193, 208, 212
John Adams 54, 122, 125, 167
John Kenneth Galbraith 186
John Locke 8, 17, 106, 107, 118, 221
John Marshall 135
Joseph Schumpeter 191
judicial fiat 20, 126, 138
judicial oversight 132

K

Kennedy 65, 66, 95, 96, 118
King George 17, 155

· 229 ·

L

league tables 162, 169
liberty 3, 6, 8, 10, 11, 13, 14, 23, 54–56, 66, 77, 79, 83, 85, 86, 99–104, 106–111, 112–114, 116–124, 129, 131, 133, 137–141, 190, 200, 201, 210, 212, 221, 226
life 8, 11, 15, 26, 53–56, 61, 63, 66, 77, 85, 102, 104, 110, 125, 137, 161, 193, 196, 217, 220
limited government 3, 13, 36, 55, 75, 77, 83, 99, 108, 118, 124, 141
limited powers 52, 69
Lincoln 65, 66, 68
litmus test 213, 215
living Constitution 138
living document 2, 12
Lord Acton 118
low quality 181

M

Magna Carta 105, 106, 122, 124
marginal tax rates 22, 94, 95, 97, 98, 100, 115
Martin Luther 105
Marx 57, 78, 80–82, 99, 219
mathematics 159, 161, 162, 165, 169, 175, 184, 224
Mayflower Compact 6
Medicaid 24, 34, 42, 84, 90, 97
Medicare 22, 24, 34, 42, 84, 90, 97
mercantilism 202, 203, 226
middle class 81, 92
military 1, 7, 16, 18, 24, 30, 31, 53, 121, 138, 142, 144–146, 148, 151, 152, 154–156, 209, 211, 212, 222
military personnel 144
Milton Friedman 186
Milwaukee Parental Choice Program 177
missile defense 146, 156
Mitch Daniels 177, 225
monotheism 59

moral absolute 51, 52
morality 2, 14, 49, 50, 52, 54–56, 58, 67, 70, 72–75

N

national debt 22, 38, 86, 87, 115
national defense 3, 6, 10, 24, 28, 37, 38, 47, 52, 63, 79, 83, 84, 140, 141, 142, 147, 148, 150, 153–156, 213
nationalized healthcare 116
national survival 215
natural rights 2, 8, 18, 19, 47, 54–60, 61–64, 67, 69, 73, 74, 103, 106, 108, 109, 111, 212
Nature's God 56, 60, 61, 219
navy 144, 145, 156
negative liberties 58, 126, 132, 133, 138, 160
Newt Gingrich 65
Newton 107
New York Times 30, 35, 218
Nuclear terrorism 153, 157

O

Obama administration 21, 25, 42, 44, 80–82, 87, 91, 97, 99, 129, 144, 146, 155, 156, 193, 208
Obamacare 18, 25, 35, 124
objectives 2, 98, 118, 129, 138, 149, 158, 159, 163, 173, 174, 176, 177, 181, 184, 202, 213
OECD 161, 165, 169, 200
One World Government 21
online learning 176, 177
On the Principles of Political Economy and Taxation 203
Origin of Species 63, 190
overcriminalization 130, 222

P

Pascal Forgione 158
Patrick Henry 101, 104
peace dividend 143, 144, 148
performance 24, 26, 44, 94, 99, 160, 162–164, 169, 170, 173, 179–181, 184, 187, 198, 213, 225
philosophers 5, 8, 103, 105
pinnacle 2, 9, 10, 16, 52, 73
PISA 160, 161, 162, 164–166, 169, 224
Plato 104, 105
Pledge of Allegiance 66, 67, 102
politically correct 167
polytheism 59, 60
positive rights 58, 132, 133, 139, 160
private workers 29, 30, 31
Progressive 13, 14, 15, 27, 55, 78, 80, 109, 110, 131–133, 137, 200, 217
property 8, 10, 46, 56, 78, 80, 81, 105, 110, 217
prosperity 79, 86, 97, 116, 151, 156, 196, 197, 201–203, 226
publick interest 189
public sector 28, 30–33, 36, 47, 116
pursuit of happiness 8, 55, 66, 102, 140

Q

quality of life 193, 196

R

raise taxes 22, 32, 33, 89, 97, 98, 100
reading 55, 159, 161, 162, 169–171, 175, 179, 180, 184, 215
redistribute 23, 186
redistribution of wealth 78
reduce spending 89, 93, 97, 100
regulatory burden 46, 99
religion 2, 7, 10, 14, 25, 49, 52–60, 56, 61–70, 72–75, 110, 160, 219
representative democracy 8, 44, 46, 47, 82

Restore the Future 2, 3, 16, 18, 19, 47, 49, 82, 88, 99, 110, 121, 139, 156, 172, 209, 211, 214, 215
revenue problem 40, 92, 93
rights of individuals 10
right-to-work 115, 129
Robert Gates 155
Roe vs. Wade 25
Ronald Reagan 20, 38, 101, 147, 191, 211
Roosevelt 13, 65, 66
rule of law 3, 52, 101, 104, 120–123, 124, 125, 127, 128, 137–139, 213

S

safety net 14, 79, 90, 201
SAT 173
school choice 176, 177, 180, 185, 225
science 13, 107, 151, 159, 161, 162, 165, 169–172, 175, 180, 184, 192, 224
Second American Revolution 2, 3, 18, 20, 99, 110, 118, 121, 128, 138, 156, 195, 209, 211, 215
Second Continental Congress 125
secular 74
separation of church and state 25, 54, 55
separation of powers 11, 24, 107, 108
social contract 6, 7, 16, 77, 107, 112, 114, 118, 119, 121, 122, 137, 138, 190, 195
Social Security 14, 22, 31, 34, 35, 42, 84, 90, 97, 147
societies 6, 7, 11, 15, 16, 50–52, 60, 64, 79, 121, 201
Socrates 104, 220
Spaulding 54, 55, 56, 111, 112, 123, 124, 219, 221, 222
spending problem 40, 92, 93
standardized tests 26, 159, 160, 169, 171, 174, 176, 184
Standard & Poors 42, 93
state and local 30, 31, 32, 76, 132, 175, 224
state government 75, 132, 160, 180

• 231 •

state of nature 5, 6, 7, 107, 112
stealing 51
students 102, 117, 128, 159–166, 168, 169, 171–181
Sun Tzu 1, 4, 215
superior being 2, 3, 55–63, 69, 70, 73, 74, 212
Supreme Court 12, 13, 20, 25, 65, 71, 125, 131, 133–135, 159
syllogism 67, 70, 74

T

tax and spend 160
tax burden 95, 96
Tax Policy Center 27
tax rates 22, 35, 36, 40, 94–98, 100, 115, 220
tax reduction 96
teachers unions 81
Tea Party 19, 82
tenure 181
terrorism 148, 149, 151, 153, 156, 157, 223
The Republic 16, 102, 104, 108, 117, 131, 133, 167, 215
Thomas Jefferson 8, 12, 17, 25, 49, 54, 60, 102, 106–109, 125, 155, 211
Thomas Paine 19, 20, 61, 125, 219
Thomas Sowell 26, 218
TIMSS 161, 164, 165, 224
transnationalism 136
transparency 43
trial by jury 126
truth-telling 50
tuition tax credit 176, 177, 185
tyranny of the majority 25, 105
tyranny of the minority 26, 67, 73, 117

U

underfunded 32
unemployment 1, 36, 37, 94, 97
unions 31, 32, 75, 80, 81, 115, 169, 180–182
United Negro College Fund 45, 158

universalist 62, 74
University of Virginia 107
USA Today 28, 29, 31, 218, 219
US Bureau of Labor Statistics 30
USC 224

V

vouchers 176, 177, 185

W

Wall Street Journal 30, 46, 194, 217, 218, 220, 223–225
Wealth of Nations 189, 190, 202, 225
well-being 46, 142, 196–198, 200, 201, 209, 213
White House Office of Management and Budget 33
Willie Sutton 92
Winston Churchill 125
Winthrop 74, 220
World Court 21

Z

zero-sum game 202, 203